BACILLUS THURINGIENSIS

CULTIVATION, APPLICATIONS IN AGRICULTURE AND ENVIRONMENTAL SAFETY

Agriculture Issues and Policies

Additional books and e-books in this series can be found
on Nova's website under the Series tab.

BACILLUS THURINGIENSIS

CULTIVATION, APPLICATIONS IN AGRICULTURE AND ENVIRONMENTAL SAFETY

DAVID P. SANDERS

EDITOR

NOTICE TO THE READER

Library of Congress Cataloging-in-Publication Data

ISBN: 978-1-53619-570-5

Published by Nova Science Publishers, Inc. † New York

CONTENTS

PREFACE

This monograph contains four chapters, each of which provides a different perspective on biopesticides. Chapter One concerns the use of biopesticides in sustainable agriculture, including the interactions between biopesticides and chemical pesticides, production issues, and opportunities for future research. Chapter Two describes the bottom-up approach for using Bacillus thuringiensis (Bt) as a biopesticide and for enhancing host plant resistance against major foliage feeders through deployment of suitable cry genes. Chapter Three reviews strategies for mitigating the spread of Bt resistance and improving insecticidal activity against Bt-susceptible pests. Lastly, Chapter Four aims to revise, debate, and evaluate the effects of Bt as phosphate solubilizing and phosphorus uptake by plant establishment.

Chapter 1 - Sustainable agriculture is on the rise in different countries of the world. On the other hand, there is an increase in the demand for food and the production of different cultures, which requires the reduction of agents that impact the environment, such as chemical inputs. This chapter will introduce biopesticides in this challenging scenario: the ones available, their market, and their interaction with chemical pesticides (as coexistence is demanded in the transition for a better production system). Furthermore, an overview will discuss the future research in prospecting for bioproducts, their production issues and potential market as effective protagonists of a sustainable agriculture in Brazil and in the world.

Chapter 2 - The wonder bacterium *Bacillus thuringiensis* (Bt), since its discovery in 1901, has been extensively studied for harnessing its potential for management of insect pests. Bt is considered the most successful microbial insecticide and occupies more than 90% share in the microbial insecticide market. In India, oilseeds are cultivated primarily in dryland areas and their cultivation is constrained by several insect pests including foliage feeders, sap sucking pests, capsule borers, etc. Research efforts have been directed towards exploitation of Bt by using the whole organism to develop sprayable formulations as well as utilizing its cry genes in genetic engineering of crops for management of the major lepidopteran pests of oilseed crops like castor, sunflower, soybean, etc. Virulent local isolates of Bt have been isolated, identified, characterized, and utilized in development of formulations for field use. Cost-effective protocols for mass production of the virulent isolates have been developed employing solid state fermentation technique for enabling and promoting indigenous Bt production with low capital investment. Efficacy of the Bt formulations against major lepidopteran pests of oilseeds has been validated through multi-location testing under the All India Co-ordinated Research Projects (AICRPs) of the Indian Council of Agricultural Research (ICAR). Data for registration has been generated in accordance with the guidelines of the Central Insecticides Board (CIB), GOI and the technologies have been licensed and commercialized to the Indian bio-pesticide industry. This article describes the bottom-up-approach being followed to use Bt as a biopesticide and also to enhance host plant resistance against major foliage feeders through deployment of suitable cry genes.

Chapter 3 - Cry toxins, one category of δ-endotoxins that are produced by the bacterium *Bacillus thuringiensis* (Bt), are used in sprays and genetically transformed crops to control insect pests. However, the rapid evolution of Bt Cry toxin resistance in target pests threatens the potency of these proteins. Thus, there is a need to elucidate the mechanism of Bt resistance and to develop strategies to cope with such resistance. Furthermore, strengthening the efficacy of Cry toxins against Bt-susceptible pests is also important. In this chapter, the authors first present a brief overview of δ-endotoxins. Next, the authors review relevant studies and discuss strategies to mitigate the spread of Bt resistance and improve insecticidal activity against Bt-susceptible pests.

Chapter 4 - *Bacillus thuringiensis* (*B. thuringiesis*) is widely known and used as a biopesticide in agriculture. It is used in the control of many crop pests. *B. thuringiensis* related studies are mostly focused on its insecticidal activity due to its entomopathogenic properties. Meanwhile, studies focusing on the biofertilizer or plant growth promoter features of *B. thuringiensis*, including its phosphorus (P) solubilizing capacity and interactions with plants, are limited. On the other hand, *B. thuringiensis* has the ability to solubilize organic and inorganic P forms in the soil and become available for plant growth. However, P is an essential nutrient for the plants and the second most important element after nitrogen. It is unavailable to plants because in the soil P is mostly present in the fixed form. *B. thruringiensis* having the phosphate solubilizing capacity, they convert the insoluble phosphate into soluble form through the changes the pH of medium, enzymes production, and organic acids and make it available for plant uptake and nutrition. It's a way and the opportunity to reduce/minimize the use of chemical fertilizers phosphate and to improve the availability of insoluble P forms in soil. In addition, the inoculation with beneficial microbes/bacteria such as *B. thruringiensis* is an ecological strategy and friendly to the environment that allows for involvement to the solubility of P in soil. Knowing the role of the *B. thuringiensis* for the crop production and environmental sustainability. The aims of this chapter are to revise, debate, and evaluate the effects of *B. thruringiensis* as phosphate solubilizing and P uptake by plant establishment.

In: *Bacillus thuringiensis*
Editor: David P. Sanders

ISBN: 978-1-53619-570-5
© 2021 Nova Science Publishers, Inc.

Chapter 1

BIOPESTICIDES IN SUSTAINABLE AGRICULTURE: CURRENT STATUS AND PROSPECTS

Diouneia Lisiane Berlitz[1,], Lidia Mariana Fiuza[2],*
Ricardo Antônio Polanczyk[3],
Daniel Mendes Pereira Ardisson-Araújo[4],
Maximiano Corrêa Cassal[5]
and Deise Maria Fontana Capalbo[6]
[1]DLB Biological Solutions, Rolante, RS, Brazil
[2]CABIO – Control Agro Bio Agricultural Research and Defense Ltda,
Porto Alegre, RS, Brazil
[3]São Paulo State University – UNESP Jaboticabal, São Paulo, Brazil
[4]Federal University of Santa Maria, Santa Maria, RS, Brazil
[5]Kyushu University, Fukuoka, Japan
[6]Embrapa Environment, Jaguariúna, São Paulo, Brazil

---

* Corresponding Author's E-mail: dberlitz@hotmail.com.

ABSTRACT

Sustainable agriculture is on the rise in different countries of the world. On the other hand, there is an increase in the demand for food and the production of different cultures, which requires the reduction of agents that impact the environment, such as chemical inputs. This chapter will introduce biopesticides in this challenging scenario: the ones available, their market, and their interaction with chemical pesticides (as coexistence is demanded in the transition for a better production system). Furthermore, an overview will discuss the future research in prospecting for bioproducts, their production issues and potential market as effective protagonists of a sustainable agriculture in Brazil and in the world.

INTRODUCTION

Agriculture is paramount for the global economic sector, given the evident increase in population and the consequent demand for food. The undeniable climatic changes affect the development of plants and are therefore a challenge for agricultural food production. Additionally, other correlated factors exist, such as demographic concentration, environmental composition (which involves other living organisms such as pests and human beings themselves), as well as economic and political aspects, and institutional, social and technological processes that make the challenges even more complex and intricate. Thus, the search for equity, sustainability and development presents significant international and intergenerational barriers and implications that cannot be overlooked (Fischer et al. 2005).

According to the FAO - Food and Agricultural Organization of the United Nations (2016), there are different challenges in the transition to sustainability that can be described into five key principles to guide strategic development:

Principle 1: Improving efficiency in the use of resources is crucial to sustainable agriculture;

Principle 2: Sustainability requires direct action to conserve, protect and enhance natural resources;

Principle 3: Agriculture that fails to protect and improve rural livelihoods and social well-being is unsustainable;

Principle 4: Sustainable agriculture must enhance the resilience of people, communities and ecosystems, especially to climate change and market volatility;

Principle 5: Good governance is essential for the sustainability of both the natural and human systems. (Text reproduced from the website: http://www.fao.org/sustainable-development-goals/overview/fao-and-the-post-2015-development-agenda/sustainable-agriculture/en/)

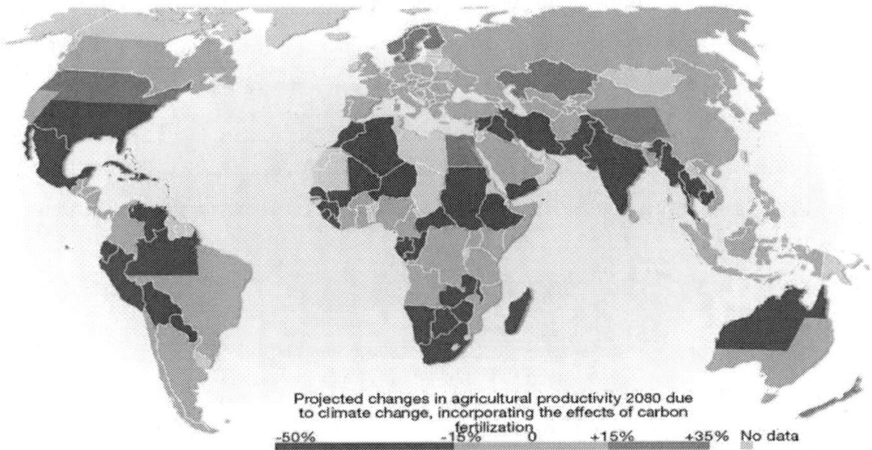

Source: https://www.grida.no/resources/6829. Cline, W.R. 2008. Global Warming and Agriculture: Impact Estimate by Country.

Figure 1. Impact of climate change on agriculture by 2080.

For the purposes of this chapter, we will focus on the most pressing issues towards agriculture. In this sense, Figure 1 shows the forecast of climate change for agriculture dealing for the next 60 years. Global warming, for example, will reduce productivity, as higher temperatures tend to accelerate the development of plants at the expense of grain production in this process. It should also interfere with the plants ability to obtain and use moisture, as there evapotranspiration is increased in warmer conditions (Cline, 2008). Different studies have focused on evaluations of agroclimatic components, simulating adaptation responses and thus offering data on opportunities for studies and scientific advances, examining options that allow farmers to minimize risks and maximize profits in conditions of climate change.

In the last few decades, human population has doubled, and food production has also increased, although not to the same extent. At the same time, the use of chemical inputs (e.g., fertilizers, pesticides) has risen due to the increase in the intensity of agricultural practices to serve this market for expanding food consumption. In the scope of sustainability, alternative forms of pest, disease and soil management are currently being sought, especially through the employment of macro and microorganisms beneficial to this habitat. The employment aims to reduce the use of inputs that cause undesirable impacts to the environment, such as chemicals with a broad spectrum of action. Among the alternative products in demand are biopesticides (here considered micro and macroorganisms with herbicidal, fungicidal, insecticidal actions and products derived from plants) (Ahirwar et al., 2019).

BIOPESTICIDES

Biological pest control is based on direct control, by using natural enemies, which tend to reduce the population of the pest and keeping it below the level of economic damage. For this control, viruses, bacteria, entomopathogenic fungi, predators and parasitoids (mostly of hymenopterans and dipterans, which live inside or outside insects to parasitize them) are used. Extracts and essential oils from several plant species are also used, in addition to semi-chemicals, such as pheromones.

A summary of those categories of biological control agents (which in this chapter we use as a synonym for biopesticides) can be found in Figure 2, which are classified according to The International Biocontrol Manufacturers' Association (IBMA).

Several control mechanisms or modes of action of the numerous biological control agents exist, which can be summarized according to Mattei (2017) as:

- antibiosis: control mechanism exerted by volatile or non-volatile compounds produced by the biological control agent, which can decrease the development speed and the production of reproductive structures of phytopathogenic fungi, controlling them.

- competition: is an occurrence in which the control agents compete for space, nutrients and/or water with the target control organisms, reducing their availability.
- parasitism: occurs with the use of hyperparasites. The control agents that have been shown to be viable are parasites of soil pathogens, such as fungi that cause root rot, wilt and seedling tipping.
- induction of resistance mechanisms: type of action attributed to control agents that have the ability to activate defense mechanisms in plants and thus compete with harmful organisms, such as phytopathogens (Polli et al. 2012). The activation of the plant metabolic routes by the control agents results in an increase in the plant resistance to pathogens, a process also called by 'induced systemic resistance'. Rhizospheric bacteria act as growth inducers, that favors the reduction of damage caused by soil pathogens. The production of siderophores and antibiotics (among other toxic products) can also be induced, which result in an interference with the pathogen recognition of the plant in its movement or growth directed at it, interfering in the pest establishment (Oostendorp and Sikora 1990; Hasky-Günther, Hoffmann-Hergarten, and Sikora 1998).

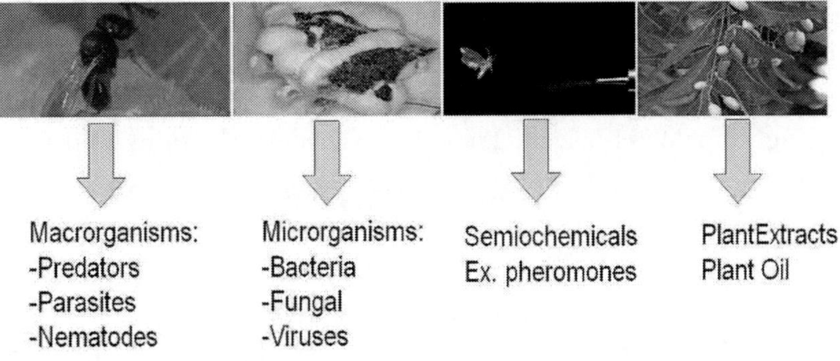

Figure 2. Categories of biopesticides. Images: https://www.koppert.com.br/podisibug/; Zambiazzi, et al., 2012; Cary, 2018; https://www.ecycle.com.br/1955-neem.html.

Microorganisms

Microbial biopesticides are products based on microorganisms (including bacteria, fungi and viruses) or their bioactive metabolites. These natural pathogens play crucial role in ecosystem balance by means of regulating host population, especially organisms emerging as a result of climate change and farming intensification.

One of the earliest mentions of the use of a pathogenic microorganism against insects was reported by V. Audouin in 1839, which described the use of a silkworm-isolated fungus for the control of insect larvae infesting trees (Rohrmann 2019). In the late XIX century, J. L. LeConti recommended the use of microorganism-causing insect diseases to control insects, and Louis Pasteur studied the use of a microsporidian parasite to control insect pests of grapes.

In a current perspective, the large-scale production of biopesticides based on insect pathogens is usually more laborious than their chemical counterparts. This may affect crop value and sometimes increase cost production. However, their use as a critical approach for the Integrated Pest Management (IPM) in a sustainable agriculture directly aggregates value to the product and may be attractive to the producers.

Bacteria

The main bacterial group of microorganisms used in the production of biopesticides are bacteria of the genus *Bacillus*. These bacteria are sporulating, found in different habitats, and often possess the peculiarity of being entomopathogenic, but may also act on other pest species not belonging to the class Insecta. Among the species belonging to this genus, *Bacillus thuringiensis*, *B. subtilis* and *B. amyloliquefaciens* have drawn great interest. These organisms have been used separately or associated with other forms of control in IPM. In the special case of *B. thuringiensis* during its sporulation phase, the bacterium produces protein inclusions with robust insecticidal (but-not-only) activity and their specificity is a consequence of the presence of *cry* genes that encode toxic proteins denominated as Cry proteins (delta-endotoxins acronym refers to the crystalline appearance under light microscopy), which possess a wide classification according to the distribution of these genes among isolates (Höfte and Whiteley 1989; Crickmore et al. 2020).

In another group of biocontrollers, *B. subtilis* can synthesize several types of antibiotics, mainly polypeptides, producing compounds such as zwittermicin A and kanosamine, lipopeptides, in addition to antifungal proteins of the bacisubin class (Bais, Fall, and Vivanco 2004; Liu et al. 2007).

The bacterium *B. amyloliquefaciens* also produces different compounds, such as surfactin, iturin A, fengycin A and fengycin B, as well as antibiotics (Leifert et al. 1995).

Another species inside the same genus, *B. licheniformis*, produces different antibiotics such as bacitracin, lantibiotics with antimicrobial capacity (Sumi et al. 2015), in addition to the production of gibberins and plant hormones (Alina, Constantinscu, and Petruţa 2015).

Considering related genera, *Lysinibacillus sphaericus* is widely used in the control of insects of order Diptera, presenting great importance for public health. Other examples are *Paenibacillus macerans* and *P. polymyxa*, which are nitrogen-fixing species capable of enhancing growth and production, also performing phosphate solubilization (V. N. da Silva 2013).

Among non-spore-forming bacteria, which can directly and indirectly contribute to crop productivity the organisms of interest, *Chromobacterium* spp. possesses toxic action against coleopterans and lepidopterans and also antimicrobial activity (Martin et al. 2007); *Pseudomonas fluorescens, P. chlororaphis, P. syringae* and *P. aureofaciens* show activity against soil fungi and seed diseases (Bettiol et al. 2012); and *Burkholderia cepacia* controls soil and seed fungi (Bettiol et al. 2012).

Fungi

One of the lines of research for pest control is the use of fungi, due to their practicality and reliability. Entomopathogenic fungi are prominent for being among the most studied forms of pest control in the world (Hayashida et al. 2014) and occupy a special status in Brazil. The pathogenicity of a fungus against insects is dependent on a complex relationship between the ability of the fungus to germinate and penetrate the cuticle and the ability of the insect immune system to prevent the growth of the fungus (Huxham, Lackie, and McCorkindale 1989).

Among entomopathogenic fungi, *Beauveria bassiana* and *Metarhizium anisopliae* stand out. These two species are arthropod pathogens widely used for their ability to infect a wide range of targets, covering most insect orders, including Lepidoptera, Coleoptera, Hymenoptera, Diptera, as well as mites (subclass Acari) (Ortiz-Urquiza, Luo, and Keyhani 2015). Additionally, they can form associations with plants, thus being able to present endophytic relationships with a wide variety of species, mobilizing nitrogen and being able to obtain carbon (in the form of various carbohydrates) from the host (Ortiz-Urquiza, Luo, and Keyhani 2015).

Among the soil fungi, the genus *Trichoderma* is highlighted due to its capacity for plant promotion linked to the solubilization and availability of macro and micronutrients for plants, through the decomposition and mineralization of plant residues, also stimulating the plant to produce root hormones. Furthermore, interactions with receptors in plants lead to the activation of signaling pathways, triggering physiological and biochemical reactions that confer induction of plant resistance to different biotic and abiotic factors (Nawrocka and Małolepsza 2013; Machado et al. 2012).

In biological control, *Trichoderma* is used as a mycoparasite of other fungi, taking nutrients from them, in addition to possessing the ability to inhibit phytopathogens through competition and production of secondary metabolites (Machado et al. 2012). These fungi are able to produce extracellular proteins, generating enzymes that degrade cellulose and chitin (Harman, Lorito, and Lynch 2004).

For the control of nematodes, the fungi *Purpureocillium lilacinum* (*Paecilomyces lilacinus*) and *Pochonia chlamydosporia* are important for having the ability to predate nematode eggs. These fungi produce serine-type lytic enzymes, proteases and chitinases, as well as exoenzymes that help the fungal mycelium penetrate nematode eggs through the partial disintegration of the vitelline layer of eggs. They can also directly inhibit embryonic development and prevent the hatching of second-stage juveniles

(Khan, Williams, and Nevalainen 2004; Bonants et al. 1995; Fernandes 2014).

Baculoviruses

Insect viruses are a great source for microbial control, especially for the control of lepidopteran defoliators (Moscardi 1999). Viruses of several different taxonomic groups are candidates as agents for pest control, including baculoviruses, cypoviruses, iflaviruses, discistroviruses, and entomopoxviruses. However, most researches focus on baculoviruses, which are important tools for agriculturally challenging insect pests, especially those on which chemical pesticides and transgenic plants approaches have failed. This group of insect viruses is the most studied around the world and largely used for both heterologous protein expression as a eukaryotic vector and agricultural and forest insect pest control (Rohrmann 2019). They are accepted as safe agents for biological control, easy for large-scale production, lethal to most of their hosts and can be formulated and used for organic and sustainable agriculture, which allows the reduction of environmental impacts from the use of other pesticides.

Importantly, new baculovirus-based biopesticides are appearing as commercialized products in many countries, especially in Brazil, and gaining an increased market share. However, the restricted number of basic studies might be a limit to the growing use of baculoviruses in the field. These viruses are characterized by a high genetic diversity in field-collected isolates, usually containing a mixture of different genotypes in variable proportions (Redman et al. 2010; Simón et al. 2004). This heterogeneity of genotypes can lead to differences in pathogenicity, mortality time, and virus production (Erlandson 2009), which are crucial traits for the use of baculoviruses as a biological insecticide (Eberle et al. 2012; Arrizubieta et al. 2015). Not only virus heterogeneity, but conditions where baculoviruses are used may also influence their efficacy to protect crops.

Baculoviruses can be used alone or combined with other insect control approaches, which make them suitable for IPM. Moreover, plant genotype can also influence the progression of baculovirus infection. For instance, insects exposed to baculoviruses during feeding on different plant species and plant exudates present different mortality rates and virus replication.

However, other biopesticide preparations can be highly competitive compared to baculoviruses, such as *B. thuringiensis* products in virtue of their simplicity of growth and formulation. Nevertheless, baculoviruses are a promising microbial tool to be used in sustainable agriculture and a growing market in the world.

For instance, the largest program of areas treated with baculoviruses took place in Brazil. AgMNPV was used to control the velvetbean caterpillar, *Anticarsia gemmatalis*, in over two million ha. The application of this virus has been undertaken since the 1970s and continued to increase until the 2000s. The success of the program was due to specific features of AgMNPV, including (1) high virulence to *A. gemmatalis* with one single application (chemicals are usually are applied twice); (2) the virus lacks genes related to the melting phenotype after death, which facilitates virus production; (3) resistance of soybeans to defoliation without reducing production (Rohrmann 2019). Interestingly, virus production was initially carried out in the fields of rural producers, in which some locations of soybeans that were naturally infested with *A. gemmatalis* were sprayed with AgMNPV and then, after 10 days, the dead carcasses hanging from the top of the plants, a characteristic of death by baculovirus infection, were collected by local workers and applied again to protect the crops. However, the demand for such viral biopesticide was increasing by the late 1990s, which led to the appearance of industries seeking to produce this baculovirus-based product in a much larger scale and quality in order to protect soybeans. Larvae were created under laboratory conditions and fed on artificial diet and their infection was managed under controlled conditions. The infected insects were subsequently processed into a wettable powder with kaolin. However, the replacement of *A. gemmatlis* by a secondary pest of the species *C. includens* as the main soybean defoliator induced a substantial retraction of AgMNPV use in the field. Other countries in South America, such as Paraguay, also have used in smaller areas this virus against the velvetbean caterpillar.

In the scope of technological innovation in 2020, two products based on *Chrysodeixis includens* multiple nucleopolyhedrovirus (ChinMNPV), indicated for the control of the caterpillar *C. includens*, were registered in Brazil. In 2019, four products were registered, all based on the following viruses: *Autographa californica* multiple nucleopolyhedrovirus (AcMNPV) + *Chrysodeixis includens* nucleopolyhedrovirus (ChinNPV) + *Helicoverpa armigera* nucleopolyhedrovirus (HearNPV) + *Spodoptera frugiperda* multiple nucleopolyhedrovirus (sfMNPV) for the control of the lepidopteran species complex *C. includens, H. armigera, S. eridania* and *S. frugiperda*. This represents a major advance in the control of the main pest lepidopterans of large crops.

Macroorganisms

Several natural enemies are used to control pests of economic importance in the group of macroorganisms (insects, mites and nematodes). The main orders of macroorganisms considered parasitoids are in the orders of Hymenoptera, Diptera and Dermaptera. Within Hymenoptera, the families most frequently employed are representatives of Braconidae and Ichneumonidae, and the Tachinidae family being the most prominent in Diptera. Other orders can be mentioned in lesser relevance, such as Strepsiptera, Coleoptera, Lepidoptera and Neuroptera. In the case of parasitoids, the insects *Cotesia flavipes* and *Trichogramma* spp. are mostly used.

Among pest predators, the families Anthocoridae, Pentatomidae, Reduviidae, Carabidae, Coccinellidae, Staphylinidae, Chrysopidae, Cecidomyiidae, Syrphidae and Formicidae are the most prominent (Silva and Britto 2015).

In addition to these insects, mites are important as biological control agents, and some bioproducts based on these organisms are already commercialized. According to Parra et al. (2002), a total of 22 orders of predators and five of parasitoids are in use.

In the phylum Nematoda, entomopathogenic nematodes (EPNs) stand out, whose genera are cosmopolitan, with examples of the recovery of EPNs in soils from six different world biomes: Neartic, Neotropical, Paleartic, Ethiopian, Eastern and Australian. This phylum includes many species, but those of economic interest have insecticidal activity and are classified within two genera: *Heterorhabditis* and *Steinernema* (Almenara et al. 2012).

Plants and their Derivatives

Global flora is characterized by its rich taxonomic diversity in different biomes among the continents. Some species originate from certain biomes and are of special interest for medicinal, agricultural, food and/or pharmaceutical use due to their metabolic constituents. Vegetable parts such as leaves, flowers, stem, bark, roots or seeds are a source of essential oils and extracts, and can be commercialized raw, fractionated or isolated according to the potential of interest. Substances from the secondary metabolism of plants can cause repellency, food and oviposition deterrence, metabolic blockades, sterilization and interference in insect development, therefore presenting insecticidal activity (Sharma and Malik 2012).

Studies with some of these products of biological origin (also called "botanicals" or "natural products") have shown action on the development of arthropods. They are options of great interest for synthetic products and usually present less damage to the environment and human health. Studies on new insecticidal molecules from plant metabolites present as an alternative of control, with some compounds such as (Sharma and Malik 2012):

(i)   Limonene and Linalool: fleas, aphids, mites, fire ants, several types of flies, paper wasps and house crickets;

(ii) Neem: varieties of sucking and chewing insects;

(iii) Pyrethrum/Pyrethrins: ants, aphids, cockroaches, fleas, flies, and ticks;

(iv) Rotenone: leaf-feeding insects, such as aphids, certain beetles (asparagus beetle, bean leaf beetle, Colorado potato beetle, cucumber beetle, flea beetle, strawberry leaf beetle, and others) and caterpillars, as well as fleas and lice on animals;

(v) Ryania: Caterpillars (European corn borer, corn earworm, and others) and thrips;

(vi) Sabadilla: Squash bugs, harlequin bugs, thrips, caterpillars, leaf hoppers, and stink bugs.

Natural neem-based insecticides (*Azadiracta indica*) are the most common products and worldwide approved for use in organic crops. This species has more than 50 terpenoid compounds, most of which act on insects. All parts of the plant possess these toxic compounds, but it is in the fruit that the highest concentration is found (Martinez, 2008-http://www.iapar.br/pagina-410.html).

Compatibility between Biopesticides Based on *Bacillus thuringiensis* and Chemical Insecticides in Tank Mixtures

The mixture of pesticides in tanks is used by farmers due to the economic infeasibility of individual application of pesticides in large areas and in order to increase the spectrum of action of products (Cloyd 2011; Gandini et al. 2020; Schreiner et al. 2016). Gazziero (2015) defined the tank mixture as the association of pesticides and equivalents in the application tank, just before spraying. Queiroz et al. (2008) established that the result of the mixture can be: additive, when the efficiency of the product is equal or similar to the individual; synergistic, when one product increases the efficiency of the other; and antagonistic, when one product negatively interferes with the efficiency of the other. The first two situations determine compatibility and the third, incompatibility (Sirvi et al. 2013).

The first recommendations by the Ministry of Agriculture in Brazil in regard to tank mixing are from 1985, when these mixtures were not prohibited, but this procedure started being practiced under the sole responsibility of the farmer, without the need for recommendation and monitoring (Lima 1997; AENDA 2011).

However, in 2017 the "Normative Instruction for Regulation of Mixtures in Tanks" became valid with more complete and accurate information (Gazziero 2015). Thereafter, tank mixing recommendations should be made by public or private institutions, of educational, research or rural extensions, producers associations, farmers cooperatives, pesticide-producing companies and corresponding organizations (IBAMA 2017). The publication of the 2017 regulation brought the need for greater compatibility studies between pesticides, whether biological or chemical, so that the farmer is correctly oriented on the tank mixture, optimizing the use of pesticides in the field..

Gazziero (2015) reports that 72% of Brazilian farmers declared to have observed problems with tank mixtures, 65% stated that the available technical information is not sufficient to guide the good agronomic practices of the mixture and, 8% revealed to be unaware of these guidelines. Abroad, Schreiner et al. (2016) observed that the majority of pesticide mixtures in Germany, France, the Netherlands and the USA include between two and five products, and may reach nine. The authors pointed out that most of the mixtures contained herbicides and their metabolites.

The State University of Northern Paraná, together with Embrapa Soybean and Agro DBO, provides compatibility tables that bring correct information to the farmer concerning the mixture of pesticide in tanks (AgroDBO 2020). This is an example of rural extension success very significant for the farmer.

Biological compatibility and its impact on pest control are more complex to assess. The effect of pesticides on entomopathogens in tank mixes may vary depending on the species/lineage of the pathogen, the chemical nature of the products and the doses used. This interaction can inhibit the vegetative growth and sporulation of microorganisms, and even cause genetic mutations. In addition, the presence of emulsifiers and other aggregates may influence compatibility with entomopathogens (Batista Filho, Almeida, and Lamas 2001).

The compatibility studies between chemicals and bioinsecticides are normally performed using *in vitro* tests of vegetative growth or survival during exposure to the chemical and are designed to mimic the types of exposure that occur in the field (Schumacher and Poehling 2012; Ribeiro et al. 2012). However, *in vitro* tests do not consider some aspects that minimize or cancel negative effects in the field, such as drift, gradual decrease in product concentration due to abiotic factors, or even irregular product deposition in the field (Rossi-Zalaf et al. 2008).

Abroad, bioinsecticide compatibility studies based on *B. thuringiensis* (here denominated as Bt insecticides) and chemical insecticides have been carried out for some decades. In a classic study by Morris (1977), the compatibility of 27 insecticides (organophosphates, carbamates, food inhibitors, pyrethrins and urea derivatives) with Thuricide® 16B, recommended for the control of lepidopteran pests, was evaluated. Among the 27 insecticides tested, only 5 were compatible with Bt insecticides, according to spore germination and vegetative cell reproduction parameters.

Karim et al. (1999) evaluated the compatibility between a Bt isolate (CEMB Bt), promising for the control of *Helicoverpa armigera* Hübner (Lepidoptera: Noctuidae), with insecticides with different modes of action. In this case, a synergistic effect was observed in the pest control when the combination of the Bt isolate with the insecticide Match® was analyzed. Regarding the same pest, Khalique and Ahmed (2001) found compatibility and synergism of the Costar® bioinsecticide (Bt *kurstaki*) and the insecticide Karate 2.5® (Lamda-cyhalothrin), under laboratory conditions.

The compatibility of chemical insecticides used in the cultivation of rice and Bt strains was analyzed by Pinto et al. (2012). The insecticides thiamethoxam, lambda-cyhalothrin, malathion and fipronil were compatible with six strains (Bt *dendrolimus*, Bt *kurstaki* HD-1, Bt *kurstaki* HD-73, Bt *thuringiensis* 144, Bt *kurstaki* ND-12 and Bt *entomocidus* 60.5) in the control of *Spodoptera frugiperda* J.E. Smith (Lepidoptera: Noctuidae), *Diatraea saccharalis* Fabr., 1794 (Lepidoptera: Crambidae), *Tibraca limbativentris* Stal, 1860 (Hemiptera: Pentatomidae), *Oryzophagus oryzae* Costa Lima (Coleoptera: Curculionalec) and *Oebalus poecilus* DeGeer 1773 (Hemiptera: Pentatomidae).

Veiga (2014) evaluated the compatibility of Bt insecticides (Dipel[®], Agree[®], and Xentari[®]) and chemical insecticides in the control of *Tuta absoluta* (Lepidoptera: Gellechidae). No bioinsecticide affected the growth and sporulation of Bt, with Tracer[®] and Azamax[®] increasing the size of bacterial colonies. Likewise, Amizadeh et al. (2015) determined that only the insecticide metaflumizone impaired the *in vitro* development of Bt *kurstaki*. The authors emphasized that the microorganism Bt can use the insecticide as a source of nutrients for its growth, corroborating with the data obtained by Veiga (2014).

Additionally, synergism and antagonism in pest control varied according to the application interval between the bioinsecticide and the insecticide; the antagonistic effect was observed when Bt *kurstaki* was applied immediately after the insecticide and when it was applied 12 or 24 hours after the insecticide application, a synergistic effect was observed in most treatments. This synergistic effect may be due to the stressful effect of the insecticide on the pest, making it more susceptible to Bt *kurstaki*.

Agostini et al. (2013) evaluated the compatibility of the bioinsecticide Dipel[®] and Agree[®] with the herbicide glyphosate in five concentrations (1.00; 1.25; 1.50; 2.00 and 2.5 L ha^{-1}). The compatibility in this case was expressed in the vegetative bacterial growth and in none of the concentrations any bacterial growth was found. Therefore, an incompatibility between the herbicide and the tested bioinsecticides was found.

Abdullah (2019) found compatibility of the Bt-based bioinsecticide (Dipel[®]) and the insecticides chlorpheniramine, cypermethrin and metomyl when evaluating colony growth *in vitro* and the control of *Tetranychus urticae* Koch (Acari: Tetranychidae) under laboratory conditions. Felland and Pitre (1990) studied a mixture of insecticides and bioinsecticides to control *Crysodeixis includens* (Lepidoptera: Noctuidae) in the USA, which in turn found synergism between Dipel[®] and the insecticides Larvin 3.2 AF[®], Lannate 1.8 L[®] and Sevil XLR 4 EC[®], in field conditions.

Navon (2000) pointed out that some insecticides may reduce food intake of the target insect, thus reducing their intake of the bioinsecticide, consequently reducing insect mortality. This fact was observed by Hoy and

Hall (1993) when they evaluated the synergism of a Bt bioinsecticide and a pyrethroid (esfenvalerate) in the control of *Plutella xylostella* (Lepidoptera: Plutellidae). On the other hand, the same author pointed out that the Bt bioinsecticide can also reduce insecticide intake, with a reduction in mortality.

Swami et al. (2012) evaluated the toxicity of 12 wettable powder formulations based on Bt *kurstaki* aiming to control *H. armigera*. Among the evaluated formulations, two containing silica precipitates were more toxic to the pest, based on the LC_{50} estimates for third instar larvae. Additionally, Gonçalves et al. (2018) reported that the formulation Dipel SC® was four times more efficient in controlling *Spodoptera albula* (Walker, 1857) (Lepidoptera: Noctuidae) than the wettable powder (PM) and dispersible granule (WG) formulations of the same bioinsecticide.

Nagy et al. (2020) researched in the literature for works that compared the toxicity of active ingredients and their formulations to the environment. The authors found 36 articles, of which 24 reported at least one observation of superior toxicity of the formulation inert material(s) in relation to the active ingredient.

Cox and Surgan (2006) pointed out that the "inert" material of a formulation may have toxic activity in its own right or may have synergistic effects. Therefore, when mixing tank products, it is important not only to consider the interactions of the active ingredients, but also of the inert ones. The interactions between different active ingredients and the inert formulas can result in a greater environmental impact of the mixtures in relation to the products alone Surgan et al. (2010). This information is important in the control of insect pests, since natural biological control is an important component of their management (Pereira et al. 2018).

According to Branscome et al. (2017), the synergistic combination of biological products with synthetic insecticides is a poorly explored concept that may be benefited with the adoption of microbial pesticides in integrated pest management. Combined, these products offer different modes of action with overlapping action spectrum to guarantee mortality, delaying the evolution of resistance in insect populations. In this sense, diamides do not exhibit contact activity, but act through ingestion with intoxication, resulting in food cessation, lethargy, paralysis and regurgitation. These products can be highly active in very low concentrations, with long residual activity that extends beyond seven days after application (Adams et al. 2016). In the case of Bt, the mode of action is by ingestion, resulting in lysis of the cells of the intestinal wall, which causes the insect to cease feeding and dying of septicemia in 2 to 3 days (Vachon, Laprade, and Schwartz 2012).

The selectivity of Bt bioinsecticides to natural enemies and beneficial non-target organisms is one of its main advantages over broad-spectrum chemical insecticides (Glare et al. 2012; Lacey et al. 2015; Rao and Jurat-Fuentes 2020). The tank mix of Bt bioinsecticides and insecticides requires to be evaluated for their possible deleterious effects of inert materials, active ingredients and their interactions in populations of non-target organisms and natural enemies (predators and parasitoids). For example, studies published in scientific journals of high impact factor have drawn attention to the effects of inert formulations on bees (Mullin et al. 2016; Fine, Cox-Foster, and Mullin 2017).

The tank mix of Bt bioinsecticides and insecticides is a complex environment due to the interaction of living microorganisms, present in bioinsecticides, chemical and inert molecules. It should be emphasized that the efficiency of a mixture of biological and chemical products against a given biological target cannot be correlated only with *in vitro* compatibility.

Global Biopesticide Market: Present and Perspectives

The world market for biopesticides is highly fragmented, with more than 40,000 companies distributed across continents. The strategy of large companies in the sector is in the launch of new products, mergers and startup acquisitions. According to the research published on the website Fortune Business Insights (2020), the global market for agricultural organic products was estimated at $ 19.73 billion in 2019, projecting values of $ 31.4 billion by the year 2027.

In the case of microbial products (bacteria, fungi and viruses), most products are formulated with bacteria, with around 68% of the world market.

In relation to pests controlled by biopesticides, insects are in first place, with 49% of control through this class of products. Despite this, compared to chemical products, the global biological market for biofungicides reaches about 12% in the marketing of products. Regarding the application technology, the foliar spray route comprises about 65% of applications today, followed by seed treatment, which already reaches 17% ("Fortune Business Insights" 2020).

The Compound Annual Growth Rate (CAGR), which is the return rate necessary for an investment to grow from its initial balance to its final balance, is considered one of the main indicators to analyze the viability of an investment. For the biopesticide market, the CAGR projected for the period 2017-2025, in different countries and continents is shown in Figure 3.

North America and Europe

North America and Europe are at the top of bioproducts commercialization, accounting for about 60% of the world market, being also the largest in production. Europe may experience a market growth, especially due to climate change that will cause an increase in temperature and the appearance of pests on the continent. A growth in the use of biofungicides in this region is also projected especially as it is considered the largest producer of fruits and vegetables, with a large presence of multinationals in the sector. In the European market for biopesticides, in addition to climatic factors, the growing demand for organic products, associated with restrictive environmental regulations, causes the sector to grow. Spain, Italy and France are the main leaders in the region's biopesticide markets ("Inkwood Research" 2020).

The reasons for high market share in North America are increased investments in research and development, well-established integrated pest management (IPM) and integrated crop management (ICM) programs.

South America and Asia Pacific

These are two emerging regions in the biopesticide market, with significant participation in the global market ("Fortune Business Insights" 2020). In South America, the increase in organic farming considered profitable and with strong export potential drives the sale and demand for bio-based products, especially microbial products.

In Asia-Pacific, increased demand for food, government initiatives to educate farmers, the growing population and increased health awareness should also lead to market growth in this region.

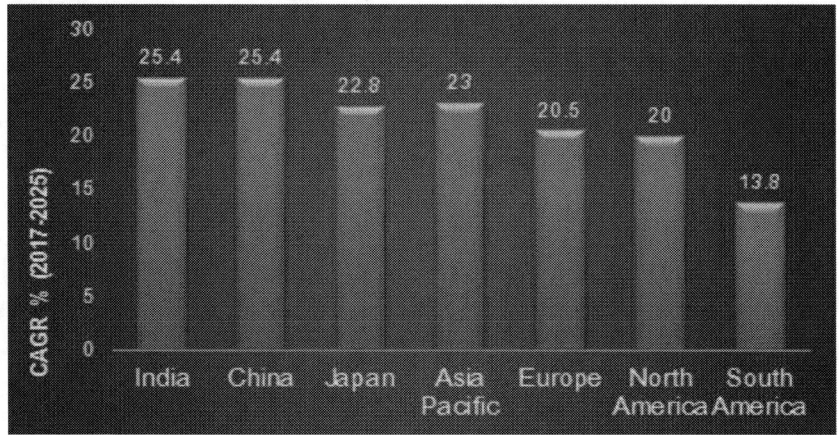

Source: Incowood Research and Fortune Business Insights (2020).

Figure 3. Compound Annual Growth Rate (CAGR) - Biopesticides Market during the forecast period 2017 - 2025.

According to Fortune Business Insights (2020), the development of countries such as India, China and Malaysia are important in the process of resilience of biopesticide companies, associated with the growing interest of large market participants in the potential of the Asian continent.

Biopesticides in Brazil: Current Overview

A considerable advance in the use of biopesticide is directly related to the increasing concern of environmental pollution and human health regarding the intensive use of chemical pesticides to control agriculture pests, among other factors. According to the Brazilian Association of Biological Control Companies (Portuguese acronym: ABCBio, 'Associação Brasileira de Empresas de Controle Biológico') the market for biological control in Brazil has increased 70%, with a financial movement over U$ 120 million in 2018 alone. Partially, the expressive increase is likely due to current status of the IPM among producers, especially to avoid selection of resistant insects to chemical insecticides and transgenic plants.

Brazil broke the record in 2020 for the registration of biological control pesticides. In a publication for the new registrations by August 20th, by the Ministry of Agriculture (Portuguese acronym: MAPA, 'Ministério da Agricultura, Pecuária e Abastecimento') in the Official Gazette, the total number for registered low-impact products was 56.

However, even being historical and increasing in demand, an important bottleneck for the use of microbial-based biopesticides relies on the fact that a pathogen and its host co-evolved in a dual balance between pathogen reproduction and disease progression. Indeed, evolution selected pathogens that often present slow speed of mortality and reduced spectrum of host species, which can substantially diminish the immediate efficacy of a specific biopesticide, allowing for more food consumption and pathogen replication. Therefore, an intensive prospection for good candidates to be used as biopesticide is of great importance for the success of a sustainable agriculture based on their use as biopesticides.

Table 1 summarizes the number of macro and microorganisms as well as plant extracts biopesticides, compared to chemicals, registered in the Brazilian Ministry of Agriculture (MAPA). Until August 2020, out of 119 products were registered for organic agriculture in Brazil, with half of them being classified as Microbiological Agents of control as insecticides, miticides or nematicides. There are currently 510 registered biopesticides in Brazil, including different macro and microorganisms, with new species such as bacteria of the genus *Pasteuria* and *Burkholderia*; seaweed extracts and essential oils as well as species of mites and entomopathogenic nematodes for control of phytonematodes, mites and especially insects.

In Brazil, the main products registered based on microbial organisms for specifically organic agriculture are fungi of two species: *Metarhizium anisopliae* and *Beauveria bassiana*. 185 products are based on these two species, as registered in the MAPA, by 32 different holding companies for the production and sale rights.

Table 1. Number of pesticide registers at the Ministry of Agriculture, Livestock, and Supply of Brazil, according to the current legislation*

Year	Biopesticides					Chemicals
	Bacteria	**Fungi**	**Viruses**	**Plants**	**Macroorganisms**	
2010	----	----	----	----	02	81
2011	01	06	----	----	08	341
2012	----	09	02	02	13	358
2013	----	07	02	01	09	268
2014	05	13	01	----	05	334
2015	19	10	10	01	12	300
2016	10	20	04	01	02	445
2017	16	14	01	----	04	448
2018	15	28	18	01	09	598
2019	16	38	08	02	14	721
2020	14	12	02	01	----	92
Sub Total	**96**	**157**	**48**	**09**	**78**	
PFAG**	**02**	**58**	**04**	**03**	**55**	
TOTAL	**98**	**215**	**52**	**12**	**133**	
Total of biopesticides: 510				**Total of chemicals: 3.986**		

* Data collected from the publications of the Federal Official Gazette, between 2010 and 2020. ** Registration for "Phytosanitary Product with approved use for Organic Agriculture - PPOA".

Other popular microorganisms are strains of *B. thuringiensis*, *B. subtilis* and *B. amyloliquefaciens*. Newer fungi species and bacteria strains have been developed against a wider range of arthropods, such as *Isaria fumosorosea* and *B. amyloliquefaciens* for use in different crops. Currently, only 52 baculoviruses, among nucleopolyhedroviruses and granuloviruses, are registered for Lepidoptera in field-grown vegetables.

In the case of macroorganisms, an analysis of the records shows that 76% of the 133 registered products are based on the parasitoid species *Cotesia flavipes* (47%), *Trichogramma galloi* and/or *T. pretiosum* (29%), possibly due to the easy and successful rearing of these organisms under controlled conditions.

In Brazil, there is a strong investment in the area of clean technologies and innovation by government agencies and private companies. Despite this, the number of biopesticides is still low when compared to the registration of chemical inputs (Figure 5). In this case, only 12% of the MAPA records from 2010 until August 2020 are biopesticides. Of these, 71% is microorganisms and 26% is macroorganisms. These data demonstrate the great potential for the discovery of new microbial species, especially through recent and increasingly accessible techniques of genetic sequencing and molecular analysis.

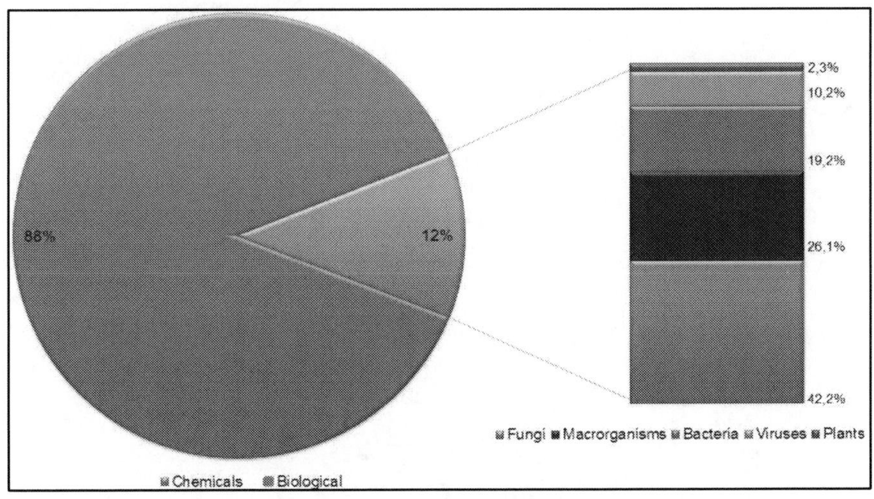

Figure 5. Data of bioproducts registered at the Ministry of Agriculture, Livestock and Supply of Brazil (MAPA), between the years 2010 and 2020.

The number of biopesticide-registering companies in Brazil is relatively large, as shown in Figure 6. It can be seen that the vast majority of these have their Headquarters in the state of São Paulo, considered the commercial center of the country. Nevertheless, industries are widely spread across the national territory.

The legislation on pesticides and equivalents (bioproducts) in Brazil is guided by the three Federal Agencies responsible for the sectors of

Agriculture (MAPA - Ministry of Agriculture, Livestock and Food Supply), Health (ANVISA - National Health Surveillance Agency) and Environment (IBAMA - Brazilian Institute for the Environment and Renewable Natural Resources). These government agencies regulate agricultural bioproducts in an effort to facilitate processes and reduce deadlines for legalizing research and registration of new biological products.

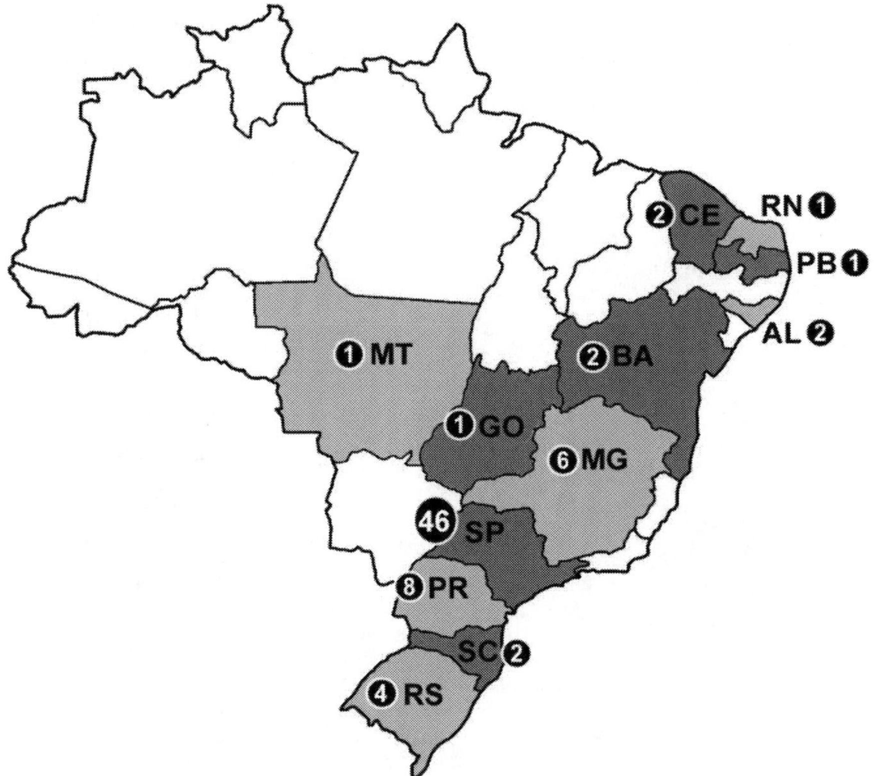

Source: Registration acts published by the Ministry of Agriculture, Livestock and Supply of Brazil (MAPA), between the years 2010 and 2020.

Figure 6. Distribution of companies that manufacture products based on macro and microorganisms and plant extracts and/or oils in the Brazilian territory.

Despite this, there is no specific legislation for biopesticides, and the registration flow follows the law of pesticides. Their components and equivalents can only be produced, handled, imported, exported, traded and used in the national territory if previously registered and met the guidelines and requirements of the Federal Agencies responsible for the agriculture, health and environment sectors, according to the organization chart shown in Figure 7.

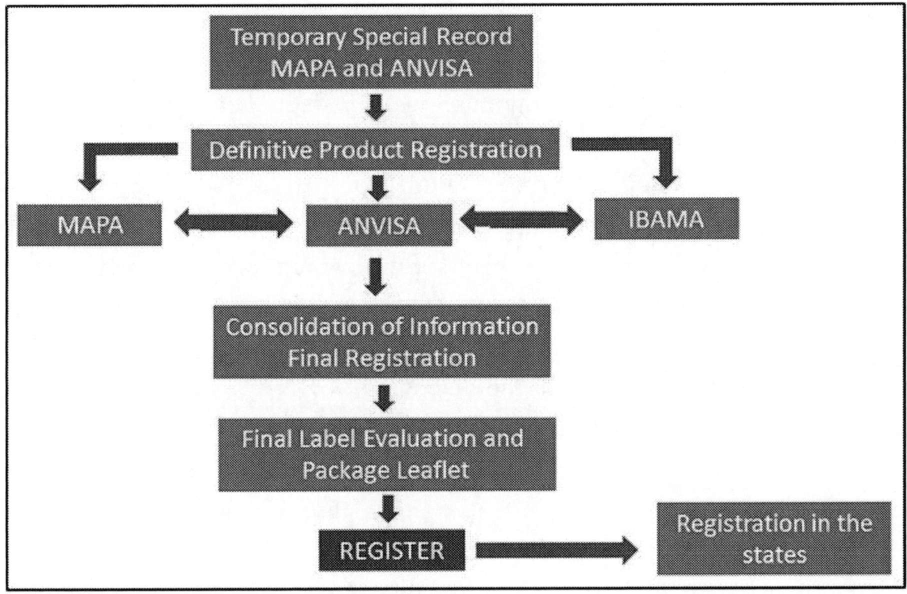

Figure 7. Registration flow of biopesticides in Brazil. MAPA - Ministry of Agriculture, Livestock and Supply; ANVISA - National Health Surveillance Agency; IBAMA - Brazilian Institute of Environment and Renewable Natural Resources.

Future Research for a Sustainable Agriculture

According to the FAO (2020), in order to become sustainable, agriculture must meet the needs of present and future generations, nourish healthy ecosystems and support sustainable management of land, water and natural resources, guaranteeing global food security.

In this sense, researches are directed towards the inclusion of different tools, molecular and biotechnological techniques to produce biopesticides in a safe and sustainable way. An example is the recombination technology of DNA and RNA, used to increase the effectiveness of biopesticides. New discoveries such as fusion proteins are being designed to develop cutting-edge biopesticides. This technology allows selected toxins (non-toxic to higher animals) to be combined with a carrier protein that makes them toxic to pests (Dhakal and Singh 2019).

In the case of antimicrobial peptides, being promising in both medicine and agriculture, studies have shown that the simultaneous use of more than one peptide increases its effectiveness in protecting plants. This is possible through the fusion of two protein sequences with the same specialized function, or complementary functions, giving rise to a new protein called fusion protein, or chimeric protein (Pinheiro et al. 2018). These authors created a "super-Blad" by fusing two proteins with distinct antimicrobial properties, a fungicide and a bactericide, which allows simultaneous treatment of diseases caused by the respective phytopathogens in agriculture.

In another line of research, the discovery of RNAi, which comprises a conserved set of mechanisms that eukaryotes use to regulate the abundance of RNA transcripts, it includes the identification of genes that have an essential physiological function in the body, through genetic targets with high transcription levels. Despite aspects such as the small acceptance of the technique in many markets and the inability to genetically transform many species, Monsanto and Dow's Smartstax maize was approved in 2017, incorporating a dsRNA against *Diabrotica virgifera virgifera* (Fletcher et al. 2020). Like any new technology, there are identified risks that must be avoided, despite the dsRNA's ability to degrade rapidly in the environment.

In the area of nanotechnology, the study of nanoscale materials (1–100 nm) called nanoparticles, which have flexible physical, chemical and biological properties, demonstrate a strong affinity for proteins (Ndolo et al. 2019). Nanotechnologies can contribute to the development of less toxic biopesticides with favorable safety profiles and greater stability of active ingredients and, together with microencapsulation, including nanoparticle-based formulations, are technologies that can accelerate the development of biopesticides (Seiber et al. 2014). Although promising, an impediment in the advances of these techniques is the size of most toxic particles of microorganisms that, many times, do not enable the micro or nanoparticularization.

However, successful cases are reported in the microencapsulation of the viral particles of the *Spodoptera frugiperda* nucleopoliedrovirus (SfNPV), whose technique protects them from inactivation by ultraviolet (UV) rays, thus presenting potential for the development of a nanoparticulate biopesticide (Ndolo et al. 2019). In the area of botanical biopesticides, research has shown that the use of nanoparticles is effective in protecting neem oil (*A. indica*) against degradation, allowing a prolonged effect on target pests (Damalas and Koutroubas 2018). With the great plant diversity found in the world, this is certainly an area of research with an endless field of possibilities to be explored.

A new area of research and application of microbiology has recently been announced: the microbiome. Berg et al. (2020) define microbiome as a characteristic microbial community, which occupies a reasonably well-defined area and that has distinct physico-chemical properties. It is the set of microbial relationships and activities that result in specific ecological niches, which form an interactive and integrated micro-ecosystem dynamics and macro-ecosystems, including eukaryotic hosts.

In this new area, the Microbiome Support (https://www.micro biomesupport.eu/) was created, including international partners such as Brazil, Canada, South Africa, China, Argentina, Australia, New Zealand, India and the USA, to improve international cooperation and the coordination of common bioeconomy research programs. The European Union's research and innovation program, to succeed Horizon 2020, is expected to invest € 100 billion in all research areas between 2021 and 2027. The representation of Brazil is through the Center for Research in Genomics Applied to Climate Change - CGACC, formed by FAPESP and the Brazilian Agricultural Research Corporation (Embrapa) at the State University of Campinas (Unicamp).

Considering the search for sustainable solutions in agro-ecosystems, it is necessary to create a sustainable environment and to implement corrective plans, covering biological solutions or even greener approaches.

Thus, themes such as "Microbes as key players" are important (Arora et al. 2018). Since microorganisms are ubiquitous and at the same time very diverse in nature, their presence everywhere suggests the roles they can play in maintaining ecosystems. Due to their flexible and adaptive genetic composition and versatile metabolic capacities, microorganisms can be used to solve various environmental issues, particularly related to the treatment of pollution, combating climate change and global warming, soil recovery, increasing agricultural productivity, facing industrial problems, remediation of waste and systems affected by pollution (Mishra, Singh, and Arora 2017; Akinsemolu 2018). In such cases, bacteria and fungi can particularly be employed to resolve issues in a simple and economical way, with minimal inputs and problems.

Another theme that approaches biopesticides is that of biodegradation, which involves the removal of pollutants usually at the source with the help of biological means, especially microorganisms. Microbial techniques are used as suitable alternatives to several traditional methods for the degradation of waste materials. A variety of microorganisms can be used to detoxify organic xenobiotic compounds in a very sustainable way. Khatoon et al. (2017) reported biodegradation as an important technique for the removal of various polymeric pollutants through microbial applications.

In bioremediation, microorganisms or other biological systems, such as plants (phytoremediation) or plants plus rhizospheric microbes (rhizorremediation) are used for the recovery of habitats already contaminated by the transformation of pollutants into less or non-hazardous substances. It is a promising, ecological and effective technology.

Microorganisms possess a great capacity to cause the remediation of chemical pesticides. Hussain et al. (2016) reviewed the degradation of several neurotoxic pesticides (neonicotinoids) in water and soil systems by applying various bacteria, such as *Bacillus, Pseudomonas, Burkholderia, Mycobacterium*, among others. In this context, Akoijam and Singh (2015) report the action of *Bacillus aerophilus* for degradation of imidacloprid, as well as Yin et al. (2012), which reports the biodegradation of cypermethrin by the bacterium *Rhodopseudomonas palustres*, and the report of the degradation of chlorpyrifos by the natural presence of *Escherichia coli* and

Pseudomonas fuorescens. According to Kumar et al. (2018) several microbial strains, such as *Aspergillus, Pseudomonas, Chlorella* and *Arthrobacter*, cause the degradation of organophosphate pesticides.

Phenolic compounds, such as bisphenols, phenols, alkylphenols and natural steroid hormones produced in various industries (including chemical pesticides) are also very dangerous for environmental health and negatively affect all organisms on Earth (Zhao, Wu, and Ma 2018). Also in these cases, scientists claim that microorganisms are very effective tools to degrade these phenolic contaminants, such as the fungus *Fusarium occiferum*, the bacteria *Acinetobacter calcoaceticus, Pseudomonas putida* and *P. stutzeri* (Mishra, Singh, and Arora 2017).

The global bioremediation market is growing at an exponential rate and has enormous potential in the future to achieve environmental sustainability goals, with greater production of healthy foods. According to the Transparency Market Research (2018), in 2016 the global bioremediation technology and services market was valued at $ 32.2 billion and is estimated at $ 65.7 billion in 2025 (much higher than projected for the microbial products - $ 31.4 billion by the year 2027).

It is interesting to note that much of the knowledge about isolation, characterization, production and application of microbials is the basis for microbial products applicable in bioremediation and biodegradation, which is a very positive point in the perspectives of current studies.

CONCLUSION

Plant protection must be considered as one of the key factors in agricultural production and biopesticides suggest to be an environmentally friendlier solution, at the same time of possessing a longer sustainability. Although there is an increasing use of biopesticides on a daily basis in order to sustain the availability of food for a growing population, the entire global farming system must aggregate biopesticides as pest control agents, by using more integrative systems and providing more assistance to farmers, especially in developing countries.

The evidences that biological technologies improve plant resilience and yields in agricultural systems are growing and are justified by the global increase in the commercialization of biopesticides. Microorganisms interact with plants and can increase the absorption and production of nutrients, improve the uptake of macro and micronutrients, control pests and mitigate plant responses to different stresses by inducing resistance. Macroorganisms help to maintain ecological balance, especially in monocultures, improving the quality attributes of agriculture.

The development of new bioproducts based on biotechnology, through genetic tools, allows the discovery of new agents and new areas that demonstrate the important role of biology in the development of sustainable agriculture with approaches favorable to the environment, agricultural production and environmental health.

In the case of Brazil, in the same way as some other countries with extensive crop production, one of the most challenging reasons to implement a sustainable agriculture is related to the intensive systems that have led to the selection of resistant insects to chemical insecticides and Bt-plants, especially of species *Spodoptera frugiperda*. The future of agriculture in these countries, particularly in Brazil, relies on substantial growth of biopesticide use and the adoption of IPM in crops and the management of insect resistance, which is already a reality in field.

Therefore, alternative methods to reduce the massive use of pesticides that cause environmental impact is fundamental to turn agriculture into sustainable. Many promising new agents are currently under development as biopesticides by Universities and Public Companies and plenty of them have been registered to be soon produced by several new start-up companies, which suggests a brighter future for biopesticides in Brazil.

REFERENCES

Abdullah, Reda. (2019). "The Side Effect of Commonly Used Chemical Pesticides on Entomopathogenic *Beauveria Bassiana* and *Bacillus Thuringiensis* as Biopesticides." *Egyptian Journal of Plant Protection*

Research Institute, 2 (1), 1–8. https://www.researchgate.net/publication/339513026.

Adams, A., Gore, J., Catchot, A.., Musser, F., Cook, D., Krishnan, N. & Irby, T. (2016). "Residual and Systemic Efficacy of Chlorantraniliprole and Flubendiamide against Corn Earworm (Lepidoptera: Noctuidae) in Soybean." *Journal of Economic Entomology*, *109* (6), 2411–17. https://doi.org/10.1093/jee/tow210.

AENDA. (2011). *Mistura Em Tanque.* São Paulo, Brazil. [*Mixing In Tank.*]

Agostini, Lucas, Karina Otuka, Elías Almeida, Mariah Valente, Lucas de Laurentis, Rogério Texeira Duarte, Thiago Trevisoli Agostini. & Ricardo Antonio Polanczyk. (2013). "Compatibilidade de Produtos à Base de *Bacillus thuringiensis* (Berliner, 1911) com Glifosato em Diferentes Dosagens, Utilizado em Soja (*Glycine max* (L.) Merrill)." *Ciencia Et Praxis*, *6*, 37–40. https://www.researchgate.net/publication/275335231.

AgroDBO. (2020). *Tabelas de Compatibilidades e Incompatibilidades Físico-Químicas de Misturas em Tanque de Agroquímicos e Fertilizantes Foliares.* 2020. https://www.portaldbo.com.br/tabela-de-compatibilidade. [*Tables of Physico-Chemical Compatibility and Incompatibilities of Tank Mixtures of Agrochemicals and Foliar Fertilizers*]

Akinsemolu, Adenike A. (2018). "The Role of Microorganisms in Achieving the Sustainable Development Goals." *Journal of Cleaner Production*, *182* (May), 139–55. https://doi.org/10.1016/j.jclepro.2018.02.081.

Akoijam, Romila. & Balwinder Singh. (2015). "Biodegradation of Imidacloprid in Sandy Loam Soil by *Bacillus aerophilus.*" *International Journal of Environmental Analytical Chemistry*, *95* (8), 730–43. https://doi.org/10.1080/03067319.2015.1055470.

Alina, Sicuia Oana, Florica Constantinscu. & Cornea Călina Petruţa. (2015). "Biodiversity of *Bacillus subtilis* Group and Beneficial Traits of *Bacillus* Species Useful in Plant Protection." *Romanian Biotechnological Letters*, *20* (5), 10737–50. http://www.bacterio.net/bacillus.html].

Almenara, D. P., Rossi, C., Neves, M. R. C. & Winter, C. E. (2012). "Nematoides Entomopatogênicos." In *Tópicos Avançados em Entomologia Molecular*, edited by M Silva Neto, C Winter, and C Termignoni, 6:56–75. São Paulo, Brazil: Instituto Nacional de Ciência e Tecnologia em Entomologia Molecular. [Entomopathogenic Nematodes. In *Advanced Topics in Molecular Entomology*]

Amizadeh, Marzieh, Mir Jalil Hejazi, Gholamreza Niknam. & Mahdi Arzanlou. (2015). "Compatibility and Interaction between *Bacillus thuringiensis* and Certain Insecticides: Perspective in Management of *Tuta absoluta* (Lepidoptera: Gelechiidae)." *Biocontrol Science and Technology*, *25* (6), 671–84. https://doi.org/10.1080/ 09583157. 2015.1007030.

Arora, Naveen Kumar, Tahmish Fatima, Isha Mishra, Maya Verma, Jitendra Mishra. & Vaibhav Mishra. (2018). "Environmental Sustainability: Challenges and Viable Solutions." *Environmental Sustainability*, *1* (4), 309–40. https://doi.org/10.1007/s42398-018-00038-w.

Arrizubieta, Maite, Oihane Simón, Trevor Williams. & Primitivo Caballeroa. (2015). "A Novel Binary Mixture of *Helicoverpa armigera* Single Nucleopolyhedrovirus Genotypic Variants Has Improved Insecticidal Characteristics for Control of Cotton Bollworms." *Applied and Environmental Microbiology*, *81* (12), 3984–93. https://doi.org/ 10.1128/AEM.00339-15.

Bais, Harsh Pal, Ray Fall. & Jorge M Vivanco. (2004). "Biocontrol of *Bacillus subtilis* against Infection of *Arabidopsis* Roots by *Pseudomonas syringae* is Facilitated by Biofilm Formation and Surfactin Production." *Plant Physiology*, *134* (1), 307–19. https://doi.org/10.1104/pp.103.028712.

Batista Filho, Antonio, José E. M. Almeida. & Clóvis Lamas. (2001). "Effect of Thiamethoxam on Entomopathogenic Microorganisms." *Neotropical Entomology*, *30* (3), 437–47. https://doi.org/10.1590/ s1519-566x2001000300017.

Berg, Gabriele, Daria Rybakova, Doreen Fischer, Tomislav Cernava, Marie Christine Champomier Vergès, Trevor Charles, Xiaoyulong Chen., et al. (2020). "Microbiome Definition Re-Visited: Old Concepts and New

Challenges." *Microbiome*, *8* (1), 1–22. https://doi.org/10.1186/s40168-020-00875-0.

Bettiol, W., Pinto, Z., Corrêa, É., Moura, A. & Bezerra, J. (2012). *Produtos Comerciais à Base de Agentes de Biocontrole de Doenças de Plantas*. Jaguariúna, Brazil. [*Commercial Products Based on Plant Disease Biocontrol Agents*]

Bonants, P. J. M., Fitters, P. F. L., Thijs, H., Den Belder, E., Waalwijk, C. & Henfling, J. W. D. M. (1995). "A Basic Serine Protease from *Paecilomyces lilacinus* with Biological Activity against *Meloidogyne hapla* Eggs." *Microbiology*, *141* (4), 775–84. https://doi.org/10.1099/13500872-141-4-775.

Branscome, D., Storey, R., Eldridge, R. & E. Brazil - US Patent, and Undefined 2017. (2017). Synergistic *Bacillus thuringiensis* subsp. *kurstaki* and chlorantraniliprole mixtures for diamondback moth, beet armyworm, sugarcane borer, soybean looper and corn earworm control. 9,723,844. *Valent BioSciences Corporation*, issued 2017. https://patents.google.com/patent/US9723844B2/en.

Cloyd, R. A. (2011). "Pesticide Mixtures." In *Pesticide Formulations*, edited by M Stoytcheva, 69–80. Rijeka, Croatia: InTech.

Cox, Caroline. & Michael Surgan. (2006). "Unidentified Inert Ingredients in Pesticides: Implications for Human and Environmental Health." *Environmental Health Perspectives*, *114* (12), 1803–6. https://doi.org/10.1289/ehp.9374.

Crickmore, Neil, Colin Berry, Suresh Panneerselvam, Ruchir Mishra, Thomas R. Connor. & Bryony C. Bonning. (2020). "A Structure-Based Nomenclature for *Bacillus thuringiensis* and Other Bacteria-Derived Pesticidal Proteins." *Journal of Invertebrate Pathology*, no. May: 107438. https://doi.org/10.1016/j.jip.2020.107438.

Damalas, Christos A. & Spyridon D. Koutroubas. (2018). "Current Status and Recent Developments in Biopesticide Use." *Agriculture (Switzerland)*, *8* (1). https://doi.org/10.3390/agriculture8010013.

Dhakal, Rishap. & Deo Narayan Singh. (2019). "Biopesticides: A Key to Sustainable Agriculture." *International Journal of Pure & Applied Bioscience*, *7* (3), 391–96. https://doi.org/10.18782/2320-7051.7034.

Eberle, K. E., Jehle, J. A., Huber, J. Integrated pest Management, and Undefined 2012. (2012). "Microbial Control of Crop Pests using Insect Viruses." In *Integrated Pest Management: Principles and Practice.*, edited by D P Abrol and U Shankar, 281–98. Oxfordshire, United Kingdom: CAB International. https://books.google.com/books?hl=en&lr=&id=0V7QBa4PIXsC&oi=fnd&pg=PA281&dq=Microbial+control+of+crop+pests+using+insect+viruses&ots=ZWC8rLjixl&sig=DE8jItBvHn2BdKzuBXLn7Wl-kdc.

Erlandson, Martin A. (2009). "Genetic Variation in Field Populations of Baculoviruses: Mechanisms for Generating Variation and Its Potential Role in Baculovirus Epizootiology." *Virologica Sinica.* https://doi.org/10.1007/s12250-009-3052-1.

FAO. (2020). *Sustainable Development Goals.* 2020. http:// www.fao.org/ sustainable-development-goals/overview/fao-and-the-post-2015-development-agenda/sustainable-agriculture/en/.

FAO, and INRA. (2016). *Innovative Markets for Sustainable Agriculture: How Innovations in Market Institutions Encourage Sustainable Agriculture in Developing Countries. Innovative Markets for Sustainable Agriculture.* http://www.fao.org/3/a-i5907e.pdf.

Felland, C. M. & Pitre, H. N. (1990). "Soybean Looper Control, Test 2, 1989." *Insecticide and Acaricide Tests.* Vol. *15.* Oxford Academic. https://doi.org/10.1093/IAT/15.1.276A.

Fernandes, Rafael Henrique. (2014). "Bionematicida à Base de *Pochonia Chlamydosporia* no Controle de *Meloidogyne Incognita* Em Cenoura." Universidade Federal de Viçosa. https://www.locus.ufv.br/handle/123456789/2033. ["Bionematicide based on *Pochonia Chlamydosporia* in the Control of *Meloidogyne Incognita* In Carrots."]

Fine, Julia D., Diana L. Cox-Foster. & Christopher A Mullin. (2017). "An Inert Pesticide Adjuvant Synergizes Viral Pathogenicity and Mortality in Honey Bee Larvae." *Scientific Reports, 7* (1), 1–9. https://doi.org/10.1038/srep40499.

Fischer, Günther, Mahendra Shah, Francesco N. Tubiello. & Harrij Van Velhuizen. (2005). "Socio-Economic and Climate Change Impacts on Agriculture: An Integrated Assessment, 1990-2080." *Philosophical*

Transactions of the Royal Society B: Biological Sciences, 360 (1463), 2067–83. https://doi.org/10.1098/rstb.2005.1744.

Fletcher, Stephen J., Philip T. Reeves, Bao Tram Hoang. & Neena Mitter. (2020). "A Perspective on RNAi-Based Biopesticides." *Frontiers in Plant Science* 11 (February). https://doi.org/10.3389/fpls.2020.00051.

"Fortune Business Insights." (2020). *Fortune Business Insights.* 2020. https://www.fortunebusinessinsights.com/industry-reports/agricultural-microbial-market-100412.

Gandini, Elizzandra Marta Martins, Elizangela Souza Pereira Costa, José Barbosa dos Santos, Marcus Alvarenga Soares, Gabriela Madureira Barroso, Juliano Miari Corrêa, Amélia Guimarães Carvalho. & José Cola Zanuncio. (2020). "Compatibility of Pesticides and/or Fertilizers in Tank Mixtures." *Journal of Cleaner Production, 268,* 122152. https://doi.org/10.1016/j.jclepro.2020.122152.

Gazziero, D. L. P. (2015). "Misturas de Agrotóxicos Em Tanque Nas Propriedades Agrícolas Do Brasil." *Planta Daninha, 33* (1), 83–92. https://doi.org/10.1590/S0100-83582015000100010.

Glare, Travis, John Caradus, Wendy Gelernter, Trevor Jackson, Nemat Keyhani, Jürgen Köhl, Pamela Marrone, Louise Morin. & Alison Stewart. (2012). "Have Biopesticides Come of Age?" *Trends in Biotechnology.* https://doi.org/10.1016/j.tibtech.2012.01.003.

Gonçalves, Kelly Cristina, Arlindo Leal Boiça Júnior, Rogerio Teixeira Duarte, Laís Fernanda Moreira, Joacir do Nascimento. & Ricardo Antônio Polanczyk. (2018). "*Spodoptera albula* Susceptibility to *Bacillus thuringiensis*-Based Biopesticides." *Journal of Invertebrate Pathology, 157,* 147–49. https://doi.org/10.1016/j.jip.2018.05.008.

Harman, G. E., Lorito, M. & Lynch, J. M. (2004). "Uses of *Trichoderma* spp. to Alleviate or Remediate Soil and Water Pollution." *Advances in Applied Microbiology, 56,* 313–30. https://doi.org/10.1016/S0065-2164(04)56010-0.

Hasky-Günther, Katja, Sabine Hoffmann-Hergarten. & Richard A. Sikora. (1998). "Resistance against the Potato Cyst Nematode *Globodera pallida* Systemically Induced by the Rhizobacteria *Agrobacterium radiobacter* (G12) and *Bacillus Sphaericus* (B43)." *Fundamental and Applied Nematology*, *21* (5), 511–17.

Hayashida, Eduardo Kenji, Samir Oliveira Kassab, Elisângela de Souza Loureiro, Camila Rossoni, Rogério Hidalgo Barbosa, Antônio de Souza Silva. & Daniele Perassa Costa. (2014). "Isolados de *Metarhizium anisopliae* (Metchnikoff) Sorokin (Hypocreales: Clavicipitaceae) Para Controle de *Diatraea saccharalis* Fabricius (Lepidoptera: Crambidae)." *EntomoBrasilis*, *7* (1), 29–32. https://doi.org/10.12741/ebrasilis. v7i1.323.

Höfte, H. & Whiteley, H. R. (1989). "Insecticidal Crystal Proteins of *Bacillus thuringiensis.*" *Microbiological Reviews*, *53* (2), 242–55.

Hoy, Casey W. & Franklin R. Hall. (1993). "Feeding Behavior of *Plutella xylostella* and *Leptinotarsa decemlineata* on Leaves Treated with *Bacillus thuringiensis* and Esfenvalerate." *Pesticide Science*, *38* (4), 335–40. https://doi.org/10.1002/ps.2780380410.

Hussain, Sarfraz, Carol J. Hartley, Madhura Shettigar. & Gunjan Pandey. (2016). "Bacterial Biodegradation of Neonicotinoid Pesticides in Soil and Water Systems." *FEMS Microbiology Letters*, no. 23, 363–252. https://doi.org/10.1093/femsle/fnw252.

Huxham, I. M., Lackie, A. M. & McCorkindale, N. J. (1989). "Inhibitory Effects of Cyclodepsipeptides, Destruxins, from the Fungus *Metarhizium anisopliae*, on Cellular Immunity in Insects." *Journal of Insect Physiology*, *35* (2), 97–105. https://doi.org/10.1016/0022-1910(89)90042-5.

IBAMA. (2017). *Portaria N° 148, de 26 de Dezembro de 2017.* Brasilia, Brazil: Ministério da agricultura, Pecuária e Abastecimento. https://www.ibama.gov.br/component/legislacao/?view=legislacao&le gislacao=137840. [*Ordinance No. 148, of December 26, 2017*]

"Inkwood Research." (2020). Inkwood Research. 2020. https://www. inkwoodresearch.com/.

Karim, S., Murtaza, M. & Riazuddin, S. (1999). "Field Evaluation of *Bacillus thuringiensis*, Insect Growth Regulators, Chemical Pesticide against *Helicoverpa armigera* (Hübner)(Lepidoptera: Noctuidae) and their Compatibility for Integrated Pest Management." *Pakistan Journal of Biological Sciences*, *2*, 320–26. http://agris.fao.org/agris-search/search.do?recordID=PK2000000012.

Khan, Alamgir, Keith L. Williams. & Helena K. M. Nevalainen. (2004). "Effects of *Paecilomyces lilacinus* Protease and Chitinase on the Eggshell Structures and Hatching of *Meloidogyne javanica* Juveniles." *Biological Control*, *31* (3), 346–52. https://doi.org/10.1016/ j.bio control.2004.07.011.

Khalik, F. & Ahmed, K. (2001). "Synergistic Interaction Between *Bacillus thuringiensis* (Berliner) and Lambda-cyhalothrin (Pyrethroid) Against, Chickpea Pod Borer, Helicoverpa armigera (Huebner)" *Pakistan Journal of Biological Sciences*. DOI: 10.3923/pjbs.2001.1120.1123.

Khatoon, Nazia, Asif Jamal. & Muhammad Ishtiaq Ali. (2017). "Polymeric Pollutant Biodegradation through Microbial Oxidoreductase: A Better Strategy to Safe Environment." *International Journal of Biological Macromolecules*. https://doi.org/10.1016/j.ijbiomac.2017.06.047.

Kumar, Shardendu, Garima Kaushik, Mohd Ashraf Dar, Surendra Nimesh, Ulrico Javier López-Chuken. & Juan Francisco Villarreal-Chiu. (2018). "Microbial Degradation of Organophosphate Pesticides: A Review." *Pedosphere*, *28* (2), 190–208. https://doi.org/10.1016/S1002-0160(18)60017-7.

Lacey, L. A., Grzywacz, D., Shapiro-Ilan, D. I., Frutos, R., Brownbridge, M. & Goettel, M. S. (2015). "Insect Pathogens as Biological Control Agents: Back to the Future." *Journal of Invertebrate Pathology*, *132*, 1–41. https://doi.org/10.1016/j.jip.2015.07.009.

Leifert, C., Li, H., Siripun Chidburee, Hampson, S., Suzanne Workman, Sigee, D., Epton, H. A. S. & Agnes Harbour. (1995). "Antibiotic Production and Biocontrol Activity by *Bacillus subtilis* CL27 and *Bacillus pumilus* CL45." *Journal of Applied Bacteriology*, *78* (2), 97–108. https://doi.org/10.1111/j.1365-2672.1995.tb02829.x.

Lima, L. C. F. (1997). "Produtos Fitossanitários: Misturas Em Tanque." *Ocepar/Coodetec/Associação Nacional de Defesa Vegetal*. Cascavel, Brazil. https://scholar.google.com/ scholar?hl=en&as_sdt=0%2C5&q= Produtos+fitossanitários%3A+misturas+em+tanque&btnG=.

["Phytosanitary Products: Tank Mixtures." Ocepar / Coodetec / National Association for Plant Defense.]

Liu, Yongfeng, Zhiyi Chen, Ng, T. B., Jie Zhang, Mingguo Zhou, Fuping Song, Fan Lu. & Youzhou Liu. (2007). "Bacisubin, an Antifungal Protein with Ribonuclease and Hemagglutinating Activities from *Bacillus subtilis* Strain B-916." *Peptides, 28* (3), 553–59. https://doi.org/10.1016/j.peptides.2006.10.009.

Machado, Daniele Franco Martins, Francini Requia Parzianello, Antonio Carlos Ferreira da Silva. & Zaida Inês Antoniolli. (2012). "*Trichoderma no Brasil: O Fungo e o Bioagente.*" *Revista de Ciências Agrárias, 35* (1), 274–88. http://www.scielo.mec.pt/ scielo.php?script= sci_arttext& pid=S0871-018X2012000100026. [Trichoderma in Brazil: The Fungus and the Bioagent. *Agricultural Sciences Journal*]

Martin, Phyllis A. W., Dawn Gundersen-Rindal, Michael Blackburn. & Jeffrey Buyer. (2007). "*Chromobacterium subtsugae* sp. nov., a Betaproteobacterium Toxic to Colorado Potato Beetle and Other Insect Pests." *International Journal of Systematic and Evolutionary Microbiology, 57* (5), 993–99. https://doi.org/10.1099/ijs.0.64611-0.

Mattei, D., Henkemeier N. P., Heling A. L., Lorenzetti, E., Kuhn, O. J. & Stangarlin, J. R. (2017). "Produtos Fitossanitários Biológicos Disponíveis para Agricultura e Perspectivas de Novos Produtos." In *Ciências Agrárias: Ética Do Cuidado, Legislação e Tecnologia Na Agropecuária*, edited by M.A. Zambom, 1st ed., 124–43. Marechal Cândido Rondon, Brazil: Universidade Estadual do Oeste do Paraná. ["Biological Phytosanitary Products Available for Agriculture and New Product Perspectives." In *Agricultural Sciences: Ethics of Care, Legislation and Technology in Agriculture*]

Mishra, Jitendra, Rachna Singh. & Naveen K. Arora. (2017). "Alleviation of Heavy Metal Stress in Plants and Remediation of Soil by Rhizosphere Microorganisms." *Frontiers in Microbiology, 8* (SEP). https://doi.org/ 10.3389/fmicb.2017.01706.

Morris, O. N. (1977). "Compatibility of 27 Chemical Insecticides with *Bacillus thuringiensis* var. *kurstaki.*" *The Canadian Entomologist, 109* (6), 855–64. https://doi.org/10.4039/Ent109855-6.

Moscardi, Flávio. (1999). "Assessment of the Application of Baculoviruses for Control of Lepidoptera." *Annual Review of Entomology, 44*, 257–89. https://doi.org/10.1146/annurev.ento.44.1.257.

Mullin, Christopher A., Julia D. Fine, Ryan D. Reynolds. & Maryann T. Frazier. (2016). "Toxicological Risks of Agrochemical Spray Adjuvants: Organosilicone Surfactants May Not Be Safe." *Frontiers in Public Health, 4* (May). https://doi.org/10.3389/fpubh.2016.00092.

Nagy, Károly, Radu Corneliu Duca, Szabolcs Lovas, Matteo Creta, Paul T. J. Scheepers, Lode Godderis. & Balázs Ádám. (2020). "Systematic Review of Comparative Studies Assessing the Toxicity of Pesticide Active Ingredients and Their Product Formulations." *Environmental Research, 181*, 108926. https://doi.org/10.1016/j.envres.2019.108926.

Navon, Amos. (2000). "*Bacillus thuringiensis* Insecticides in Crop Protection - Reality and Prospects." *Crop Protection, 19* (8–10), 669–76. https://doi.org/10.1016/S0261-2194(00)00089-2.

Nawrocka, J. & Małolepsza, U. (2013). "Diversity in Plant Systemic Resistance Induced by *Trichoderma*." *Biological Control* 67: 149–56. https://www.sciencedirect.com/science/article/pii/S1049964413001539 ?casa_token=4e_zdGOwlgwAAAAA:8R8H2JGYQBCHw93XcAMu OoroM4U_fjmyg9GRCnBliY4tz3sx1J-E0wJuvQ1h0rgBlx-Kyrc.

Ndolo, Dennis, Elizabeth Njuguna, Charles Oluwaseun Adetunji, Chioma Harbor, Arielle Rowe, Alana Den Breeyen, Jeyabalan Sangeetha., et al. (2019). "Research and Development of Biopesticides: Challenges and Prospects." *Outlooks on Pest Management, 30* (6), 267–76. https://doi.org/10.1564/v30_dec_08.

Oostendorp, M. & Sikora, R. (1990). "*In-Vitro* Interrelationships between Rhizosphere Bacteria and *Heterodera schachtii*." *Revue de Nématologie, 13* (3), 269–74.

Ortiz-Urquiza, Almudena, Zhibing Luo. & Nemat O. Keyhani. (2015). "Improving Mycoinsecticides for Insect Biological Control." *Applied Microbiology and Biotechnology*. Springer Verlag. https://doi.org/ 10.1007/s00253-014-6270-x.

Parra, J. R., Botelho, P. S. M., Corrêa-Ferreira, S. & Bento, J. M. S. eds. (2002). "Comercialização de Inimigos Naturais no Brasil: Uma Área

Emergente." In *Controle Biológico No Brasil, Parasitóides e Predadores*, 343–49. São Paulo, Brazil: Manole. https://scholar. google.com/scholar?hl=en&as_sdt=0%2C5&q=Comercialização+de+i nimigos+naturais+no+Brasil%3A+uma+área+emergente&btnG=. [Commercialization of Natural Enemies in Brazil: An Emerging Area. In *Biological Control In Brazil, Parasitoids and Predators*]

Pereira, R. R., Neves, D. V. C., Campos, J. N., Santana, P. A., Hunt, T. E. & Picanço, M. C. (2018). "Natural Biological Control of *Chrysodeixis includens*." *Bulletin of Entomological Research*, *108* (6), 831–42. https://doi.org/10.1017/S000748531800007X.

Pinheiro, Ana Margarida, Alexandra Carreira, Ricardo B. Ferreira. & Sara Monteiro. (2018). "Fusion Proteins towards Fungi and Bacteria in Plant Protection." *Microbiology (United Kingdom)*, *164* (1), 11–19. https://doi.org/10.1099/mic.0.000592.

Pinto, Laura M. N., Natália C. Dörr, Ana Paula A. Ribeiro, Silvia M. de Salles, Jaime V. de Oliveira, Valmir G. Menezes. & Lidia M. Fiuza. (2012). "*Bacillus thuringiensis* Monogenic Strains: Screening and Interactions with Insecticides Used against Rice Pests." *Brazilian Journal of Microbiology*, *43* (2), 618–26. https://doi.org/10.1590/ S1517-83822012000200025.

Polli, Anderson, Andrea Neves, Fabiana Gallo, Janaina Gazarini, Sandro Rhoden. & João Pamphile. (2012). "Aspectos Da Interação Dos Microrganismos Endofíticos Com Plantas Hospedeiras e Sua Aplicação no Controle Biológico de Pragas na Agricultura." *SaBios-Revista de Saúde e Biologia*, *7* (2), 82–89. [Aspects of the interaction of endophytic microorganisms with host plants and their application in biological pest control in agriculture. *SaBios-Journal of Health and Biology*]

Queiroz, Angélica Araujo., Martins, J. A. J. & Cinha, J. P. R. A. (2008). "Adjuvantes e Qualidade Da Água Na Aplicação de Agrotóxicos." *Bioscience Journal*, *24* (4), 8–19. http://www.seer.ufu.br/index.php/ biosciencejournal/article/download/6923/4587. [Adjuvants and Water Quality in the Application of Pesticides. *Bioscience Journal*]

Rao, Tejas. & Juan Luis Jurat-Fuentes. (2020). "Advances in the Use of Entomopathogenic Bacteria/Microbial Control Agents (MCAs) as

Biopesticides in Suppressing Crop Insect Pests." In *Biopesticides for Sustainable Agriculture*, edited by N Birch and T Glare, 99–134. Cambridge, UK: Burleigh Dodds Science Publishing. https://doi.org/10.19103/as.2020.0073.06.

Redman, Elizabeth M., Kenneth Wilson, David Grzywacz. & Jenny S. Cory. (2010). "High Levels of Genetic Diversity in *Spodoptera exempta* NPV from Tanzania." *Journal of Invertebrate Pathology, 105* (2), 190–93. https://doi.org/10.1016/j.jip.2010.06.008.

Ribeiro, L. P., Blume, E., Bogorni, P. C., Dequech, S. T. B., Brand, S. C. & Junges, E. (2012). "Compatibility of *Beauveria Bassiana* Commercial Isolate with Botanical Insecticides Utilized in Organic Crops in Southern Brazil." *Biological Agriculture and Horticulture, 28* (4), 223–40. https://doi.org/10.1080/01448765.2012.735088.

Rohrmann, George F. (2019). *Baculovirus Molecular Biology - NCBI Bookshelf.* National Center for Biotechnology Information. 2019. https://www.ncbi.nlm.nih.gov/books/NBK49500/.

Rossi-Zalaf, L. S., Alves, S. B., Lopes, R. B., Neto, Silveira, S. & Tanzini, M. R. (2008). "Interação de Micro-Organismo com Outros Agentes de Controle de Pragas e Doenças." In *Controle Microbiano de Pragas na América Latina: Avanços e Desafios*, edited by S B Alves and R B Lopes, 279–302. Piracicaba, Brazil: FEALQ.

Schreiner, Verena C., Eduard Szöcs, Avit Kumar Bhowmik, Martina G. Vijver. & Ralf B. Schäfer. (2016). "Pesticide Mixtures in Streams of Several European Countries and the USA." *Science of the Total Environment, 573*, 680–89. https://doi.org/10.1016/ j.scitotenv.2016. 08.163.

Schumacher, Verona. & Hans Michael Poehling. (2012). "*In Vitro* Effect of Pesticides on the Germination, Vegetative Growth, and Conidial Production of Two Strains of *Metarhizium anisopliae*." *Fungal Biology, 116* (1), 121–32. https://doi.org/10.1016/ j.funbio.2011.10.007.

Seiber, James N., Joel Coats, Stephen O. Duke. & Aaron D. Gross. (2014). "Biopesticides: State of the Art and Future Opportunities." *Journal of Agricultural and Food Chemistry, 62* (48), 11613–19. https://doi.org/10.1021/jf504252n.

Sharma, Shilpi. & Promila Malik. (2012). "Review Article Biopestcides: Types and Applications." *International Journal of Advances in Pharmacy, 1* (4), 508–15.

Silva, Aldemi B. & Britto, J. M. (2015). "Controle Biológico de Insetos-Pragas e Suas Perspectivas." *Revista Agropecuária Técnica, 36* (1), 248–58. https://scholar.google.com/scholar?hl=en&as_sdt= 0%2C5 &q=Controle+biológico+de+insetos-pragas+e+suas+perspectivas +para+o+futuro&btnG=. [Biological Control of Insect-Pests and Their Perspectives. *Technical Agricultural Magazine*]

Silva, Valéria Nogueira da. (2013). *Interação de Microrganismos Na Solubilização de Fósforo e Potássio de Rochas para Produção de Biofertilizantes.* Natal, Brazil: Universidade Federal do Rio Grande do Norte. https://repositorio.ufrn.br/jspui/handle/123456789/12652.

Simón, Oihane, Trevor Williams, Miguel López-Ferber. & Primitivo Caballero. (2004). "Genetic Structure of a *Spodoptera frugiperda* Nucleopolyhedrovirus Population: High Prevalence of Deletion Genotypes." *Applied and Environmental Microbiology, 70* (9), 5579–88. https://doi.org/10.1128/AEM.70.9.5579-5588.2004.

Sirvi, S. L., Choudhary, H. R., Jat, N., Tiwari, V. K. & Singh, N. (2013). "Compatibility of Bio-Agents with Chemical Pesticides: An Innovative Approach in Insect-Pest Management." *Popular Kheti, 1,* 62–67. https://scholar.google.com/scholar?hl=en&as_sdt=0%2C5&q=Compat ibility+of+bio-agents+with+chemical+pesticides%3A+an +innovative+approach+in+insect-pest+management&btnG=.

Sumi, Chandra Datta, Byung Wook Yang, In Cheol Yeo. & Young Tae Hahm. (2015). "Antimicrobial Peptides of the Genus *Bacillus*: A New Era for Antibiotics." *Canadian Journal of Microbiology, 61* (2), 93–103. https://doi.org/10.1139/cjm-2014-0613.

Surgan, Michael, Madison Condon. & Caroline Cox. (2010). "Pesticide Risk Indicators: Unidentified Inert Ingredients Compromise Their Integrity and Utility." *Environmental Management.* https://doi.org/ 10.1007 /s00267-009-9382-9.

Swami, Dinesh, Bishwajeet Paul, & Dotasara, S. K. (2012). "Effect of Carriers on the Efficacy of *Bacillus thuringiensis* var. *kurstaki* (HD-1)

Formulations." *Journal of Biological Control*, *26* (3), 245–50. https://doi.org/10.18311/jbc/2012/3496.

Transparency Market Research. (2018). *Bioremediation Technology & Services Market to Expand at a CAGR of ~7% during 2020 to 2030.* Transparency Market Research. 2018. https:// www.transparencymarket research.com/pressrelease/bioremediation-technology-services-market.htm.

Vachon, Vincent, Raynald Laprade. & Jean Louis Schwartz. (2012). "Current Models of the Mode of Action of *Bacillus thuringiensis* Insecticidal Crystal Proteins: A Critical Review." *Journal of Invertebrate Pathology*, *111* (1), 1–12. https://doi.org/10.1016/ j.jip.2012.05.001.

Veiga, A. C. P. (2014). "Compatibilidade Entre Produtos Químicos e Biológicos à Base de *Bacillus thuringiensis* Berliner no Controle de *Tuta absoluta* (Meyrick) (Lepidoptera: Gelechiidae)." São Paulo, Brazil: Universidade Estadual Paulista Júlio de Mesquita Filho. https://repositorio.unesp.br/handle/11449/115916. [*Compatibility Between Chemical and Biological Products Based on Bacillus thuringiensis Berliner in the Control of Tuta absoluta (Meyrick) (Lepidoptera: Gelechiidae).*]

Yin, Lebin, Xinghua Li, Yong Liu, Deyong Zhang, Songbai Zhang. & Xiangwen Luo. (2012). "Biodegradation of Cypermethrin by *Rhodopseudomonas palustris* GJ-22 Isolated from Activated Sludge." *Fresenius Environmental Bulletin*, *21* (2A), 397–405. https://www. researchgate.net/publication/287904383.

Zhao, Lin, Qi Wu. & Aijin Ma. (2018). "Biodegradation of Phenolic Contaminants: Current Status and Perspectives." In *IOP Conference Series: Earth and Environmental Science*, *111*, 12024. Institute of Physics Publishing. https://doi.org/10.1088/1755-1315/111/1/012024.

In: *Bacillus thuringiensis*
Editor: David P. Sanders

ISBN: 978-1-53619-570-5
© 2021 Nova Science Publishers, Inc.

Chapter 2

HARNESSING THE POTENTIAL BENEFITS OF *BACILLUS THURINGIENSIS* FOR MANAGEMENT OF INSECT PESTS OF CASTOR (*RICINUS COMMUNIS L.*)

P. S. Vimala Devi, P. Duraimurugan and M. Sujatha[*]
ICAR-Indian Institute of Oilseeds Research,
Hyderabad, Telangana State, India

ABSTRACT

The wonder bacterium *Bacillus thuringiensis* (Bt), since its discovery in 1901, has been extensively studied for harnessing its potential for management of insect pests. Bt is considered the most successful microbial insecticide and occupies more than 90% share in the microbial insecticide market. In India, oilseeds are cultivated primarily in dryland areas and their cultivation is constrained by several insect pests including foliage feeders, sap sucking pests, capsule borers, etc. Research efforts have been directed towards exploitation of Bt by using the whole organism to develop sprayable formulations as well as utilizing its cry genes in genetic engineering of crops for management of the major lepidopteran pests of

---

[*]Corresponding Author's E-mail: mulpuri.sujata@icar.gov.in

oilseed crops like castor, sunflower, soybean, etc. Virulent local isolates of Bt have been isolated, identified, characterized, and utilized in development of formulations for field use. Cost-effective protocols for mass production of the virulent isolates have been developed employing solid state fermentation technique for enabling and promoting indigenous Bt production with low capital investment. Efficacy of the Bt formulations against major lepidopteran pests of oilseeds has been validated through multi-location testing under the All India Co-ordinated Research Projects (AICRPs) of the Indian Council of Agricultural Research (ICAR). Data for registration has been generated in accordance with the guidelines of the Central Insecticides Board (CIB), GOI and the technologies have been licensed and commercialized to the Indian bio-pesticide industry. This article describes the bottom-up-approach being followed to use Bt as a biopesticide and also to enhance host plant resistance against major foliage feeders through deployment of suitable cry genes.

Keywords: *Bacillus thuringiensis, cry* genes, genetic engineering, castor, pest management

INTRODUCTION

Agriculture along with its allied sectors constitutes the primary occupation and livelihood for 58% of Indian population and contributes to 15.4% India's Gross Domestic Product. Limited land resources, population increase and increased demand for food necessitate minimizing the crop losses due to insect pests. On-farm yield losses range from 10-30% while losses in oilseeds alone during harvesting, handling and storage are around 10% (Varaprasad and Duraimurugan, 2016). Management of several insect pests with microbial biopesticides based on fungi, bacteria, viruses and nematodes or their bioactive compounds has been gaining momentum in India due to their reduced environmental toxicity, target specificity and safety to non-target organisms (Kumar et al., 2019).

Bacillus thuringiensis (Bt) is the most popular and successful insect pathogen with over a century of history as a potent biocontrol agent for management of insect pests of agriculture, forest ecosystems and human vectors. Bt is a ubiquitous, gram positive sporulating soil bacterium that forms insecticidal crystal proteins during sporulation phase of its growth cycle which cause specific toxicity in insects. Bt was first reported by Ishiwata during 1901 in Japan and named as *Bacillus sotto*. Ernst Berliner rediscovered it ten years later in 1911 in the province of Thuringen, Germany, from a diseased caterpillar of flour moth and classified it as type species *Bacillus thuringiensis*. The proteinaceous crystal production within the sporangium was detected by Hannay in 1953 who proved that the crystal protein was toxic to insects. This toxin was named as δ-endotoxin by Heimpel in 1967. The first commercial Bt product Sporeine became available in France during 1938. In the 1950s, widespread use of biocontrols began to take hold in USA as a host of research on Bt efficacy was published. More than 818 Bt toxins have been reported world-wide. There is still a continuous search for novel Cry proteins with toxic potential against various insect pests. Bt accounts for about 5–8% of *Bacillus* spp. population in the nature. Currently, the commercial utilization of Bt includes marketable pesticidal formulations as well as development of insect-resistant genetically modified crops employing the genes encoding the insecticidal toxins of this bacterium making several million dollars in global market. This has also resulted in significant lowering of usage of chemical pesticides coupled with the increase and survival of beneficial and non-target organisms.

Widespread adoption of cheaper but more toxic synthetic chemical insecticides in the second half of 20th century kept the research and development on Bt at a low ebb. An industrial process known as submerged fermentation, which allowed production of Bt on a large scale was developed by the Pacific Yeast Product Company that gave a fillip to development and application of new products of Bt, principally in niche markets where petroleum-based chemicals were not registered, ineffective, or uneconomical.

Harnessing the potential of Bt for effective pest management necessitates identification of virulent isolates, development of protocols for their cost-effective production along with suitable formulations with an extended shelf-life at ambient temperatures. Extensive studies have been undertaken world over for detecting the presence of Bt in diverse ecological habitats *viz.*, soil, stored product dust, dead insects, food grains, phyllosphere, and aquatic environments. A vast distribution and diversity of

Bt strains bearing putative genes with specific toxicity spectrum is reported from various habitats of India including rice fields ecosystems, Andaman and Nicobar Islands, desert soils of Rajasthan, phylloplanes of crops grown in Delhi, Western Ghats of Karnataka and Tamilnadu, Hill zone soils of Karnataka and as a whole from Southern, Northern states and North eastern states of India (Kaur and Singh, 2000; Asokan and Puttaswamy, 2007; Ramalakshmi and Udayasuriyan, 2010, Vimala Devi and Vineela, 2016; Subbannaet al., 2017). However, the efforts in India have been minimal reporting primarily new isolates of Bt as well as effective isolates against some insect pests like *Helicoverpa armigera* and *Spodoptera litura* through laboratory bioassays (Lalitha et al., 2012; Lone et al., 2017).

Castor (*Ricinus communis* L.) (Malpighiales: Euphorbiaceae) is an industrially important non-edible oilseed crop and its oil is used in the manufacture of several industrial products like lubricants, paints, plasticizers, pharmaceuticals, cosmetics, soaps, biopolymers, biodiesel, etc. (Ogunniyi, 2006; Shrirame et al., 2011). India, Mozambique, China and Brazil are the major castor growing countries in the world (http://www.fao.org/faostat/). India accounts for 79% and 87% of the world's castor area and production, respectively meeting about 90% of the world's castor oil requirement. Though castor productivity in India is more than world average, there are several production constraints. One of the major constraints is the excessive damage caused by insect pests. Among them lepidopteran foliage feeders *viz.*, semilooper (*Achaea janata* L.), tobacco caterpillar (*Spodoptera litura* F.) and capsule borer (*Conogethes punctiferalis* Guen.) are of greater economic importance. It is estimated that castor yields are reduced by 17.2 to 63.3% due to the insect pests (Lakshminarayana and Duraimurugan, 2014). The ICAR-Indian Institute of Oilseeds Research (ICAR-IIOR), Hyderabad, India adopted a participatory bottom-up approach. Accordingly, research was initiated wherein the major problem of crop damage due to foliage feeders under rainfed cultivation of castor by resource poor farmers were addressed in Bt technology development right from identification of the effective strain to mass production through solid-state fermentation (SSF) and product development for commercial use through formulations of virulent Bt strains as well as genetically engineering with toxin coding genes. We present hereunder an overview of the efforts embarked upon and the outcome thereof towards exploiting Bt for the management of lepidopteran insect pests in castor.

DEVELOPMENT OF BT AS A BIOPESTICIDE

Development of Sprayable Formulations of Bt

Research efforts for identification of virulent Bt isolates were initiated at ICAR-IIOR during 1998 with collection of soil samples from top soils at 5 cm depth from cultivated and fallow fields from Mahabubnagar, Nalgonda and Karimnagar districts of Telangana; Guntur and Bapatla districts of Andhra Pradesh; Nagpur in Maharashtra; Dharwad in Karnataka and various districts of Rajasthan. Selective isolations by the sodium acetate method of Travers et al. (1987) yielded around 200 Bt isolates. These isolates were multiplied in nutrient broth and the resulting Bt powders were screened through laboratory bioassays against major lepidopteran pests namely *H. armigera, A. janata, S. litura, S. exigua* and *S. frugiperda* and the potent isolates were identified (Vimala Devi et al., 2001; 2005; Vimala Devi and Vineela, 2016). These isolates were subjected to molecular characterization through PCR using gene specific primers as well as rep-PCR (Reddy et al., 2012a, b; Vimala Devi and Vineela, 2016). The next step was development of low-cost mass production protocols. Bt is an aerobic bacterium with specific growth requirements of pH 7.2-7.5 and temperature 30 ^{0}C for optimal growth, sporulation and toxin production. Hence, Bt production has been in the domain of multinational companies and traditionally multiplied through submerged fermentation requiring high capital investment thereby resulting in high cost of production. However, technology adoption in a developing country like India necessitates development of protocols that make the mass production economically feasible, so that these new technologies can compete in the open market.

Biopesticides production in India is being promoted commercially by medium range entrepreneurs. However, Bt production was not undertaken by them due to high capital investment for production as well as the difficulty in generating data for registration. In order to bring Bt in the reach of the average Indian farmer, a novel, simple, low-cost mass production methodology for Bt on the principle of solid-state fermentation (SSF), eliminating the need for a fermentor, was developed at ICAR-IIOR during the year 2002. One local isolate DOR Bt-1 belonging to *B. thuringiensis* var. *kurstaki*, isolated from a dead castor semilooper larva collected from farmers' fields at Kothakota, Mahabubnagar district of Telangana, was identified for the mass production. This process of mass production of Bt was carried out utilizing easily available and inexpensive media ingredients (Vimala Devi et al., 2005). Bt obtained through this process was formulated as a wettable powder and used for further studies. The process does not pose contamination problems and can be carried out effectively in a routine manner even by less skilled/unskilled people. All the growth requirements were addressed in the development of the production process through SSF. This was the first report of enabling Bt production through solid-state fermentation, without employing a fermenter, using agricultural wastes/byproducts like wheat bran and molasses. This methodology had the potential to enable large-scale localized production of Bt through establishment of cottage industry/micro-enterprises (Vimala Devi and Rao, 2005a; b). Based on this protocol, other virulent isolates were also multiplied against important pests.

Formulations developed at ICAR-IIOR include wettable powder (WP) formulations of two Bt strains DOR Bt-1 and 5, a suspension concentration (SC) formulation of DOR Bt-127, a water dispersible granule (WDG) formulation of Bt-127, two combination SC formulations of Bt with the entomopathogenic fungi *Beauveria bassiana* and *Nomuraea rileyi* (Vimala Devi et al., 2020; 2021).

Evaluation of Efficacy

All the formulations were evaluated through laboratory bioassays and field trials at ICAR-IIOR as well as farmers' fields (Vimala Devi and

Sudhakar, 2006; Vimala Devi et al., 2020). Data on LC_{50} and potency was generated against *H. armigera, S. litura* and *A. janata*. These formulations were then evaluated through All India Co-ordinated Research Projects (AICRPs) against *A. janata* and *S. litura* on castor, *H. armigera* and *S. litura* on sunflower, *S. litura* on soybean as well as *H. armigera* on pigeonpea. DOR Bt-1 formulation was found to be effective for control of castor semilooper @ 1.0 g/l, @ 2.0 g/l against pod borer in pigeon pea and head borer in sunflower while DOR Bt-127 SC formulation was found effective against *S. litura* and semiloopers in soybean as well as against head borer in sunflower @ 3.0 ml/l in multilocation trials under AICRPs and was found superior to the commercial Bt formulation Delfin WG that is overly expensive at Rs. 4000/- per kg.

Outreach Initiatives

ICAR-IIOR took the initiative of reaching out to farmers for creative awareness about the importance of the technology and facilitates its adoption by them. Outreach activities at Kothakota and Nallavelli villages included on-farm training programmes about the use and efficacy of Bt against lepidopteran larvae, its mode of action and safety to the natural enemies *viz.*, parasitiods and predators. The trainings included identification of the various stages of the pests right from egg to the late larval instars as well as the various parasitoids. Emphasis was laid on undertaking the Bt sprayings when the larvae were in the early stages.

Another outreach initiative was the conduct of demonstrations with the formulations in farmers' fields. Each farmer was provided the IIOR Bt formulation for one acre and another acre was maintained as per his/her practice for pest management. Thus, farmers were made to observe the differences in between Bt sprayed fields and insecticide sprayed fields *viz.*, natural enemies, frequency of sprays, yield etc. (Vimala Devi and Sudhakar, 2006).

The SSF technology for Bt production had the potential to enable large-scale localized production of Bt through establishment of cottage industry/micro-enterprises. Hence, two microenterprises were established in association with NGOs - one with Society for Development of Drought Prone Area (SDDPA) and second with Grameena Mahila Mandali (GMM) at Mahabubnagar and Nalgonda districts, respectively. High school dropout boys and girls were trained in Bt production and the formulation was supplied to farmers for use in field for pest management. This initiative helped in faster spread of technology among the farmers and its successful adoption.

Intellectual Property Rights (IPR)

Intellectual property (IP) refers to creations of the mind, such as inventions; literary and artistic works; designs; and symbols, names and images used in commerce. The IP Rights usually give the creator an exclusive right over the use of his/her creation for a certain period. It was for the first time that a protocol was developed for Bt production through SSF. Since the Bt SSF technology had a commercial value, IP protection became utmost important. A provisional patent application for the process was filed by ICAR in 2002 and the complete patent application in 2003.

Development of stable formulations of Bt in combination with entomopathogenic fungi by IIOR was the first report. A patent has been granted for the process (Indian patent no. 315134). Bt and entomopathogenic fungi have diverse modes of action thereby creating more stress on the target insect since Bt acts through per ingestion while fungi infect the insect through the cuticle. This in turn leads to faster kill of the larvae and effective against older larvae as well.

Registration with Central Insecticides Board (CIB)

The import, manufacture, sale, transport, distribution and use of microbial pesticides in India is regulated under the Insecticides Act, 1968 and rules framed there under. Registration of bio-pesticides in India was approved around the year 1998. Biopesticides of botanical and microbial origin were included in the schedule of the Insecticide Act 1968.

Since micro-enterprises were established for enabling localized production and sale of the DOR Bt-1 W.P. formulation to the farmers in Mahabubnagar and Nalgonda districts of Telangana, registration of the formulation became essential. DOR Bt-1 formulation was therefore registered provisionally with the CIB in 2005 under the trade name KNOCK.W.P. [Registration No. CIR–511/2005(256)] under section 9(3b) with the Central Insecticides Board, Govt. of India, and was the first formulation of a microbial pesticide from the Indian Council of Agricultural Research, registered for commercial use and also the first from public sector.

Technology Licensing

Medium range entrepreneurs promote commercial production of biopesticides in India. However, the high capital investment for Bt production as well as the difficulty in generating data for registration did not enable commercial production and promotion of Bt. Data generation by the IIOR for the purpose of registration encouraged several medium range firms to approach IIOR and request for licensing of the technology during the year 2003. IIOR started licensing the Bt production and formulation technology to the bio-pesticide industry in India from July 1, 2006. The technology package included DOR Bt-1 strain along with data for provisional registration and training in the production technology through SSF. The technology has been licensed to more than 40 biopesticide firms in India. This initiative enabled the firms to seek registration of the formulation with the Central Insecticides Board and undertake its commercial production. Majority of the firms who received Bt-1 license, obtained the provisional registration and sold their Bt formulations under different trade names *viz.*, Cezar, Caterpillin, JAS BT, VBT, R.B. Bt, Prasar, Dipole, Beater etc. Companies have now started receiving permanent registration. Thus, DOR Bt technology has successfully reached the end user *i.e.*, the average Indian farmer and is being used in insect pest management on several crops.

Access and Benefit Sharing

ICAR-IIOR was the first institute to set an example in access and benefit sharing in accordance with the Biodiversity Act, 2014. IIOR shared 3% of the licensing fee with Biodiversity Management Committee (BMC) at Kothakota, Mahabubnagar district through the Telangana State Biodiversity Board (TSBB). The BMC was constituted by TSBB at Kothakota since the DOR Bt-1 isolate was obtained from that village. BMCs have the responsibility of using these funds for betterment of the respective village including creation of awareness on conservation of biodiversity. For this initiative, IIOR received the prestigious UNDP award as runner up under the category "Successful Mechanisms/Models for Access and Benefit Sharing" at the "India Biodiversity Awards 2016."

GENETIC ENGINEERING THROUGH DEPLOYMENT OF Bt*CRY* GENES

One of the strategies for genetic enhancement of resistance to insect-pests is development of transgenic plants through incorporation of suitable insect resistance candidate genes (Sharma et al., 2000). Transgenics are imperative for agriculturally important crops and the 113-fold hectare increase between 1996 and 2018 makes biotech crops the fastest adopted crop technology in agriculture. Of the global biotech area of 192 million ha, genetically engineered (GE) varieties expressing one or several insecticidalgenes from Bt were grown on a total of 101 million hectares worldwide,reaching adoption levels above 80% in some regions (ISAAA, 2017; 2018; Romeiset al., 2019). With the advent of genomics, several genes in the fatty acid metabolic pathways and triacyl glycerol assembly; genes controlling the noxious proteins ricin and RCA and disease resistance genes have been characterized. Hence, there is vast scope for genetic engineering of castor so as to mitigate the damages caused by the lepidopteran insect-pests and improve castor for seed quality traits.

Identification of Candidate Bt Proteins Effective Against the Foliage Feeders

Host plant resistance is the most desirable option for insect control. Development of transgenics for resistance against lepidopteran insect pests would facilitate the development of genotypes resistant to these biotic stresses through breeding programmes. Due to narrow genetic resources conferring resistance to insect pests in castor, there is a need for utilizing the biotechnological tools for alien gene transfer. Before embarking on the programme of genetic engineering, it is vital to identify suitable insect resistance genes, which could be deployed into castor against the major pests. Several candidate genes, such as crystal protein (Cry) genes of Bt produced during the sporulation stage, vegetative insecticidal Bt proteins (VIPs) induced during the vegetative stage, proteinase inhibitors, lectins, α-amylase inhibitors, insect chitinases, novel genes of plant origin, etc., are deployed into crop plants for imparting protection against insect-pest. However, the most commonly used and commercially exploited insect resistance genes are the Bt*cry* genes from Bt.

Insecticidal δ-endotoxins of Bt have acquired great significance because of their specificity to target pests, non-toxicity to humans and beneficial insects, toxicity at low concentration and environmental friendly nature. It is known that Btvar*kurstaki* (Btk) strains produce several lepidopteran toxic proteins such as Cry 1Aa, Cry 1Ab, Cry 1Ac, Cry IIAandCry 1B. Information on the reaction of the major lepidopteran pests attacking castor to Cry proteins in the toxin specificity database (http://wmv. glfc.forestry.ca/bacillus/web98.adb) is limited. Studies were carried out at ICAR-IIOR to assess the efficacy of various purified crystal Bt proteins which are lepidopteran pest specific against major defoliator pests of castor. Bioassays were done against neonate larvae of semilooper (*A. janata*), tobacco caterpillar (*S. litura*), hairy caterpillars (*Spilosoma obliqua* and *Euproctis fraterna*) using Cry toxins (Cry1Aa, 3A, 2B, 1C, 2A, 1E, 1Ac, 1F, 9A, 1Ab) at concentrations ranging from 4 to 1500 ng/cm^2 using leaf paint assay. With regard to semilooper, the proteins 1Aa, 1Ab, 1E and 2A were most effective, resulting in 100% within 48 hours while the other proteins gave no or delayed mortality at the highest concentration tested. Among the effective proteins, Cry1Aa was found to be superior to other proteins in giving early mortality even at lower concentrations (125 ng/cm^2) (Sujatha and Lakshminarayana, 2005). In case of tobacco caterpillar, none of the proteins gave 100% mortality even 96 hours after treatment at the highest concentration (1500 ng/cm^2) tested except for Cry1Aa which gave 50% mortality at 1500 ng/cm^2. Increasing the concentration of the proteins up to 3000 ng/cm^2 also failed to cause larval mortality. However, feeding cessation in terms of low larval weight was recorded in treatments with Cry1Aa and 1Ab (Lakshminarayana and Sujatha, 2005). The proteins 1Aa, 1E, 1Ab were effective against Bihar hairy caterpillar, *S. obliqua* while 1Ac, 1Aa were effective against hairy caterpillar, *E. fraterna*. Based on these studies, the genes *cry1Aa, cry1Ab, cry1Ac* and their fusion genes were identified for deployment into castor for conferring resistance to the foliage feeders.

The genes deployed in castor were *cry1AcF* (Manoj Kumar et al., 2011); *cry1Aa, cry 1Ec* (Sujatha et al., 2009) and *cry1Ab* (Malathi et al., 2006). The *cry1EC* and *cry1AcF* genes are deployed for conferring resistance to *S. litura*, while *cry1Aa* and *cry1Ab* were used for *A. janata*. Keeping in view the superiority of Cry1Aa protein against all the target pests, genetic transformation of castor has been initiated using the respective gene with plant codon usage. *S. litura*, a polyphagous lepidopteran insect pest on castor is reported to be tolerant to most of the known δ-endotoxin proteins. Hence, a hybrid δ-endotoxin protein of Cry1Ea and Cry1Ca was developed by replacing amino acid residues 530-587 in a poorly active natural Cry1Ea protein with a highly homologous 70 amino acid region of Cry1Ca in domain III and was designated as Cry1Ec (Singh et al., 2004). The solubilised Cry1Ec made from *E. coli* was 4-fold more toxic to the larvae than Cry1Ca, the best known δ-endotoxin against *Spodoptera* sp. The hybrid endotoxin conferred complete protection against *S. litura* when deployed in tobacco and cotton. Based on these studies, the genes *cry1Aa, cry 1Ab* and *cry 1Ec* were selected to be deployed in castor for conferring protection against the major lepidopteran foliage feeders.

Tissue Culture and Regeneration in Castor

Availability of an efficient and highly reproducible system of tissue culture regeneration is a prerequisite for genetic transformation experiments. Castor proved to be highly recalcitrant to manipulations *in vitro* which is a major bottleneck in the development of transgenic castor (Sujatha et al., 2008). Earlier studies during 1960s were confined to endosperm culture and the interest was mainly due to the large endospermic seeds that enabled easy culturability. However, these resulted in continuously growing cultures which lacked the ability for organogenic differentiation. Tissue culture studies were once again undertaken by researchers during 1980s for obtaining whole plantlet regeneration from seedling tissues of castor. Experiments were restricted to the use of young seedlings and the ability to regenerate complete plants was rather limited. Plant regeneration was mainly from the pre-existing meristematic centers (Athma and Reddy, 1983; Reddy and Bahadur, 1989; Molina and Schobert, 1995; Alamet al., 2010), and a maximum of 40 and 47 shoots from embryo axes and shoot tip explants, respectively, was reported (Sujatha and Reddy, 1998). Callus-mediated shoot regeneration from hypocotyl explants, young stem segments, leaves and cotyledonary leaves are reported but the morphogenic differentiation was sporadic, unreproducible with very low frequency of shoot regeneration and few shoots (1-5) per responding explants (Reddy et al., 1987; Genyu, 1988; Bahadur et al., 1992; Sarvesh et al., 1992). Subsequently, Ahn et al., (2007), Sujatha and Reddy (2007) and Ganesh Kumari et al., (2008) have reported relatively higher shoot induction frequencies from seedling explants with around 22 to 24 shoots per explant and the use of growth adjuvants and amino acids for improving the caulogenic ability. Preincubation of cotyledon explants from mature seeds cultured on medium with 5 µM thidiazuron (TDZ) in dark for seven days resulted in a maximum of 25 shoots per explant (Ahn and Chen, 2008). Nevertheless, the reproducibility of these methods across laboratories and different genotypes is rather limited.

Owing to the caulogenic response of cultured explants of castor, efforts were made at refinement of protocols for meristem based proliferation. Of the meristematic explants tested, embryo axes possessed high proliferative ability on TDZ supplemented media when compared to shoot apices and nodal explants (Sujatha and Reddy, 1998). This protocol using mature seed derived embryo axes was used for genetic transformation of castor (Sujatha and Sailaja, 2005; Malathi et al., 2006). To unravel the reasons for recalcitrance of castor tissues, transcriptomic profiles of cultured hypocotyl tissues of castor were compared with that of jatropha and sunflower which have great prosperity for regeneration (Sai Sudha et al., 2019). Differential gene expression analysis indicated downregulation of genes involved in auxin biosynthesis and homeostasis, vacuolar transporter genes, Wuschel gene, shoot root like transcription factors and histidine containing phosphotransfer proteins while genes like DELLA and brassinosteroid LRR receptor kinases were upregulated resulting in regeneration recalcitrance. There is a need for assessment of a large number of genotype and growth regulator combinations for determining the caulogenic ability and also to draw valid conclusions about the recalcitrance of castor tissues *in vitro*. Till such time a highly efficient and reproducible system of plant regeneration is developed, genetic transformation experiments have to rely on meristem-based shoot proliferation system.

Development of Transformation Protocols

Castor leaves are susceptible to crown gall disease caused by *Agrobacterium tumefaciens* (Lippincott and Haberlein, 1965). However, recalcitrance *in vitro* has been a major problem for undertaking plant transformation experiments in castor. McKeon and Chen (2003) obtained genetically engineered plants by employing the method of *Agrobacterium*-mediated transformation through vacuum infiltration of wounded flower buds (US Patent No 6.620.986). The first successful attempt to develop a stable transformation system for castor using embryo axes from mature seeds has been described by Sujatha and Sailaja (2005). The protocol exploits the proliferative ability of embryo axes on TDZ (0.5 mg/l) supplemented media followed by selection on appropriate antibiotic multiplication and elongation on media supplemented with BA (0.5 and 0.2 mg/l). Elongated shoots were rooted on a media fortified with 200 mg/l NAA. A similar shoot proliferation method with minor modifications has been adopted for direct gene transfer using particle gun bombardment method (Sailajaet al., 2008). Malathi et al., (2006) developed *Agrobacterium* mediated method while Kumar et al., (2011) optimized *in planta* method of transformation for deployment of *cry* genes. The several transformation studies for genetic transformation of castor continue to rely on meristem-based proliferation (Chen et al., 2013; Patel et al., 2015; Sousa et al., 2017).

Development of Transgenic Events

The vector-mediated and direct gene transfer methods were employed for transformation of castor, DCS-9 (Jyoti) using appropriate vectors containing the Bt fusion gene *cry1EC* driven by enhanced 35S promoter (Sujatha and Sailaja 2005; Sailaja et al.,2008; Sujatha et al.,2009). About 93 putative transformants were regenerated following selection on hygromycin and kanamycin. The integration, inheritance and expression of the introduced genes was demonstrated up to T_4 generation by PCR, Southern analysis and ELISA analysis. Field bioassays against *S.litura* and *A. janata*, conducted for eight events in T_1 to T_4 generations under net confinement conditions resulted identification of promising events bestowed with resistance to the two major defoliators. The same procedure with minor modifications was used for production of semilooper resistant transgenic castor by incorporating a synthetic δ-endotoxin *cry1Ab* gene driven by CaMV 35S promoter which recorded feeding inhibition, reduced larval growth and 88.9 to 97.3% mortality of *A. janata* as compared to the untransformed control plants (13.9%) (Malathi et al., 2006). The construct harbouring the insect resistance gene carried the herbicide resistance gene (*bar*) for selection of putative transformants. The presence of introduced gene, its stable integration, expression and inheritance was confirmed through PCR, Southern analysis, ELISA and progeny tests. However, characterization of transgenics harbouring the *cry1Ab* was confined to the analysis of T₁ generation plants, and their reaction to the target pest was evaluated under laboratory. Transgenic castor with *cry1AbcF* gene against *S. litura* resulted 22.7 to 96.6% mortality (Kumar et al., 2011).

Bioassays and Field Evaluation for Event Selection

Among the four studies where transgenic events were developed, event selection was carried out with stable transgenic lines harboring the *cry1Aa* and *cry1Ec* genes for imparting insect resistance. Events harboring *cry1Ab* and *cry1ACF* genes were subjected to laboratory bioassays against *A. janata* and S. *litura*, respectively, while the events harboring *cry1Aa* and *cry1Ec* genes were tested for their level of resistance to the two foliage feeders both under laboratory and field conditions. Characterization of the transgenic castor events with the *cry1Aa* gene against lepidopteran insect pests in laboratory bioassays revealed that the mortality of *S. litura*and *A. Janata* ranged from 20 to 80% in different transgenic events and the weight reduction of surviving larvae over the control larvae after 8 days of feeding was 28.4–87.2% in the case of *S. litura* and 27.9–78.1% for *A. Janata* (Tarakeswari et al., 2019). In different events carrying the *cry 1Ec* gene, average mortality in laboratory bioassays against *S. litura* was 20–40% and 30–70% against *A. janata*. Reduction in larval weight of the surviving larvae at 8 days after treatment was 20–70% in case of *S. litura* and 20–80% for *A. janata* (Sujatha et al., 2009). For assessing the reaction in the field, heavy incidence of both the major foliage feeders was ensured by encaging the experimental plots with nets and encouraging natural built up of *A. Janata* and through artificial releases of *S. litura*. Under such high level of pest load of both the pests, the untransformed control (DCS-9) was completely defoliated including spike damage, while the events harboring the *cry* genes (*1Aa* and *1Ec*) showed defoliation less than 25% leaf damage. Transgenic events displayed significant variations with regard to the level of defoliation based on the crop growth stage, generation in which they were subjected to the bioassays which was substantiated through protein expression analysis using ELISA. Events harboring *cry1Ec* gene *viz.*, NBRI-PB-1, PCP 202-

AMT-1 and PCP 202-AMT-9 and with *cry 1Aa* genes *viz.,* AMT-894, AK1304-PB-1, AK1304-PB-4, and DTS-43) were found to be promising. As castor can tolerate defoliation up to 25%, the genes could be transferred to other agronomically superior genotypes or stacked together and tested for the bioefficacy of the introduced gene(s) in different genetic backgrounds.

Biosafety Concerns

Over the past 20 years, extensive laboratory and field-based studies of the non-target effects of crops producing Cry proteins revealed that insecticidal proteins used in commercialized Bt crops cause no direct, adverse effects on non-target species outside the order (i.e., Lepidoptera for Cry1and Cry2 proteins) or the family (i.e., Coleoptera, Chrysomelidae for Cry3 proteins) of the target pest(s). This also holds true for Bt plants with pyramided insecticidal proteins (Romeis et al., 2019). Besides, when Bt crops replace synthetic chemical insecticides for target pest control, this creates an environment supportive of the conservation of natural enemies. In respect to Bt-transgenic crops, the National Academies of Sciences, Engineering, and Medicine (NASEM, 2016) concluded:

> "On the basis of the available data, the committee found that planting of Bt crops has tended to result in higher insect biodiversity on farms than planting similar varieties without the Bt trait that were treated with synthetic insecticides." Earlier the European Academies have stated that "There is compelling evidence that GM crops can contribute to sustainable development goals with benefits to farmers, consumers, the environment and the economy" (EASAC, 2013).

Following event selection, it is essential to determine the gene flow, protein toxicity and safety to beneficial organisms. In the case of castor, the main concerns are the toxicity of castor seeds due to ricin, an extremely toxic and water-soluble ribosome-inactivating protein which is also present in lower concentrations in other parts of the plant posing a problem in determining the toxicity and allergenicity of the transgenic castor events. The *Bt* genes for introduction in castor were selected based on the initial bioassays of the purified proteins against the target lepidopteran pests. Risk assessment studies necessitate testing of the transgenic events/cry proteins on the non-target and beneficial organisms. Eri silkworm (*Samia cynthia ricini*) is an important economic insect used in the production of valuable silk commonly known as 'vanya silk.' The silkworm primarily reared on castor leaves and the process is called eri culture. When assayed, purified Bt crystal proteins (Cry1Aa, Cry1Ab, Cry1Ac) showed toxicity to eri silkworm. Among ten purified crystal proteins of Bt tested at concentrations ranging from 2.93 to 3000 ng/cm^2 for their toxicity to eri silkworm through protein paint bioassays using castor leaves, Cry1Aa (2.6 ng/cm^2) was highly toxic followed by Cry1Ac (29.3 ng/cm^2) and Cry1Ab (68.7 ng/cm^2). The Cry1Ca and Cry1Ea proteins exhibited moderate toxicity to eri silkworm larvae and resulted in 23% and 28% mortality, respectively at the highest concentration tested (3000 ng/cm^2) while other tested proteins were with low/non-toxic (Kumar et al., 2016). Evaluation of transgenic castor events (AK1304-PB-1, AK1304-PB-4 and AMT-894) expressing the insecticidal protein Cry1Aa against eri silkworm exhibited 20.2 to 78.5% reduction in larval weight (Tarakeswari et al., 2019). Hence, it represents the importance of long term perspectives and technical amenability of economic part of transgenic plant, for biosafety assay, in meeting the regulatory requirement for the successful and ecofriendly commercialization of transgenic crops. Further, studies need to be taken up for identification of proteins that are toxic to the target pests of castor but not attacking eri silkworm for development of genetically modified castor.

CONCLUSION AND FUTURE PERSPECTIVES

The indiscriminate uses of chemical pesticides have caused incalculable damage to every aspect of the environment. As a result, agriculturists world over are persuading farmers to turn to ecologically sound pest management technologies that do not harm the environment. Bt based biopesticides are promising alternatives to chemical pesticides and have opened up new vistas in insect pest management for safe and eco-friendly pest management.

Despite its high target specificity and environmentally favorable "green" characteristics, environmental sensitivity (photolysis and rain fastness) and short persistent effect reduces its potency and duration of efficacy against pests. In the recent years, nanotechnology has been used to improve the performance and safety of the microbial pesticides and to reduce the dosage needed in comparison to conventional formulations. Hence, improved formulations of the active ingredient are essential to make Bt formulations comparable to chemical insecticides in terms of the speed of kill. Formulations with small and narrow particle-size distribution will improve the coverage of sprays on the foliage. However, the requirements for generation of toxicology data need to be developed to enable registration and commercial exploitation for effective crop protection.

Host plant resistance is the most important aspect in the management of castor pests. Genetic engineering confers resistance in plants to biotic stresses and increases yield in several crops. The introduction of foreign genes by genetic engineering techniques requires an efficient *in vitro* regeneration system for the desired plant species. Such a system must be rapid, reliable and applicable to a broad range of genotypes. Difficulty in tissue culture-based regeneration and poor reproducibility of results is the major bottleneck for genetic transformation of castor. Despite the research expanded over the past four decades in tissue culture, a genotype-independent direct or callus-mediated adventitious shoot regeneration system remains the much sought-after goal in castor. There is an immediate need for development of a highly efficient, reliable and reproducible direct and callus-mediated tissue culture system as a prelude for genetic engineering of castor for desirable traits. Efficacy of crystal protein genes from Bt and other proteins against another dreaded castor pest *viz.*, capsule borer (*Conogethes punctiferalis*) has to be evaluated for enhancing host plant resistance to all the major lepidopteran pests in castor. Potential impacts on biodiversity, non-target organisms, soil microbiota and toxicological studies for oil (crude and refined) on rat/mice need to be studied in detail from safety point of view, before their deliberate releases into the environment. This study represents a typical and unique example of using the Bt both as a biopesticide and also as a transgenic event and presents a comprehensive account of the prospects and problems with each of the technologies and also at various development efforts to bring the products to a commercial stage.

REFERENCES

Ahn Y J and Chen G Q, "*In vitro* Regeneration of Castor (*Ricinus communis* L.) using Cotyledon Explants," *Hort Science* 43, (2008): 215-219.

Ahn Y J, Vang L, McKeon T, As, and Chen G Q, "High-frequency Plant Regeneration through Adventitious Shoot Formation in Castor (*Ricinus communis* L.)," *In Vitro Cellular and Developmental Biology Plant* 43, (2007): 9-15.

Alarn I, Shamin S A, Mondal S C, Alam M J, Khalekuzzaman M, Anisuzzaman M, and Alam M F, "*In vitro* Micropropagation through Cotyledonary Node Culture of Castor Bean (*Ricinus communis* L.)," *Australian Journal of Crop Science* 4, (2010): 81-84.

Asokan R and Puttaswamy, "Isolation and Characterization of *Bacillus thuringiensis* Berliner from Soil, Leaf, Seed Dust and Insect Cadaver," *Journal of Biological Control* 21, no. 1 (2007): 83-90.

Athma P and Reddy T P, "Efficiency of Callus Initiation and Direct Regeneration from Different Explants of Castor (*Ricinus communis* L.)," *Current Science* 52, (1983): 256-257.

Bahadur B, Reddy K R K, and Rao G P, "Regeneration Potential of Callus Cultures in Castor (*Ricinus communis* L.)," *Asian Journal of Plant Sciences* 4, (1992): 13-18.

Chen Y, Wang Y, Wang Y, Huang F, Li G and Wang W, "Construction of RNAi Binary Vector of Ricin a Chain and its Transformation," *Acta Botanica Boreali-Occidentalia Sinica*33, no. 1 (2013): 39-42.

EASAC, *Planting the Future: Opportunities and Challenges for using Crop Genetic Improvement Technologies for Sustainable Agriculture.* European Academies Science Advisory Council, EASAC Policy Report 21, (2013), Available online: <https://easac.eu/publications/

details/planting-the-future-opportunities-and-challenges-for-using-crop-genetic-improv ement-technologies-fo/>.

Ganesh Kumari K, Ganesan M, and Jayabalan N, "Somatic Embryogenesis and Plant Regeneration in *Ricinus communis*," *Biologia Plantarum* 52, (2008): 17-25.

Genyu Z, "Callus Formation and Plant Regeneration from Young Stem Segments of *Ricinus communis* L.," In: *Genetic Manipulation in Crops,* IRRI, Artillery House, Artillery Row, London, United Kingdom (1988), pp. 393.

ISAAA, *Global Status of Commercialized Biotech/GM Crops in 2017: Biotech Crop Adoption Surges as Economic Benefits Accumulate in 22 Years*, ISAAA Brief No. 53, (2017), ISAAA: Ithaca, NY.

ISAAA, *Global Status of Commercialized Biotech/GM Crops in 2018*, ISAAA Brief No. 54, (2018), ISAAA: Ithaca, NY.

Kaur S and Singh A, "Distribution of *Bacillus thuringiensis* Isolates in Different Soil types from North India," *Indian Journal of Ecology* 27, (2000): 52-60.

Kumar A M, Rohini S, Kalpana N R, Ganesh P T, and Udayakumar M, "Amenability of Castor to an *Agrobacterium*-Mediated *in planta* Transformation Strategy Using a *cry1AcF* Gene for Insect Tolerance," *Journal of Crop Science and Biotechnology* 14, no. 2 (2011): 125-132.

Kumar K K, Sridhar J, Murali-Baskaran R K, Senthil-Nathan S, Kaushal P, Dara S K, and Arthurs S, "Microbial Biopesticides for Insect Pest Management in India: Current Status and Future Prospects," *Journal of Invertebrate Pathology* 165, (2019): 74-81, https://doi.org/10.1016/ j.jip.2018.10.008.

Lakshminarayana M and Duraimurugan P, "Assessment of Avoidable Yield Losses due to Insect Pests in Castor (*Ricinus communis* L.)," *Journal of Oilseeds Research* 31, no. 2 (2014): 140-144.

Lakshminarayana M and Sujatha M, "Toxicity of *Bacillus thuringiensis* var. *kurstaki* Strains and Purified Crystal Proteins against *Spodoptera litura* (Fabr.) on Castor, (*Ricinus communis* (L.)," *Journal of Oilseeds Research* 22, no. 2 (2005): 433-434.

Lalitha C, Muralikrishna T, and Chalam M S V, "Isolation, Identification, Bioassay and Field Evaluation of Native *Bacillus thuringiensis* Strains against *Spodoptera litura* (Fabricius) in Groundnut (*Arachis hypogaea*)," *Journal of Biological Control* 26, no.1 (2012): 34-42.

Lippincott J A and Haberlin G T, 1965 "The Induction of Leaf Tumours by *Agrobacterium tumefaciens,"American Journal of Botany* 52, (1965): 396-403.

Lone A S, Malik A, and Padaria J C, "Selection and Characterization of *Bacillus thuringiensis* Strains from Northwestern Himalayas Toxic against *Helicoverpa armigera*," *Microbiology Open Wiley*, (2017), https://doi.org/10.1002/mbo3.484.

Malathi B, Ramesh S, Rao K V, and Reddy V D, "*Agrobacterium*-mediated genetic transformation and production of semilooper resistant transgenic castor (*Ricinus communis*L.)," *Euphytica*147, (2006): 441-449.

McKeon T A and Chen G Q, "Transformation of *Ricinus communis*, the Castor Plant," (2003) US Patent No. 6.620.986.

Molina S M and Schobert C, "Micropropagation of *Ricinus communis,"Journal of Plant Physiology* 147, (1995): 270-272.

NASEM, *Genetically Engineered Crops: Experiences and Prospects, National Academies of Sciences*, (2016), Engineering, and Medicine, The National Academies Press, Washington, DC, USA.

Ogunniyi D S, "Castor Oil: a Vital Industrial Raw Material," *Bioresource Technology* 97, (2006): 1086-1091.

Patel M K, Joshi M, Mishra A, and Jha B, Ectopic Expression of SbNHX1 gene in Transgenic Castor (*Ricinus communis* L.) Enhances Salt Stress by Modulating Physiological Process. *Plant Cell, Tissue and Organ Culture* 122 (2015): 477-490.

Ramalakshmi A and Udayasuriyan V, "Diversity of *Bacillus thuringiensis* Isolated from Western Ghatsof Tamil Nadu State, India," *Current Microbiology* 61, (2010): 13-18. https://doi.org/10.1007/s00284-009-9569-6.

Romeis J, Steven E N, Michael M, and Anthony M S, "Genetically Engineered Crops help Support Conservation Biological Control," *Biological Control* 130, (2019): 136-154.

Reddy K R K and Bahadur B, 1989 "Adventitious Bud Formation from Leaf Cultures of Castor (*Ricinus communis* L.)," *Current Science* 58, (1989): 152-154.

Reddy V P, Narasimha Rao N, Vimala Devi P S, Lakshmi Narasu M, Dinesh Kumar, V, 2012a "PCR-Based Detection of *cry* Genes in Local *Bacillus thuringiensis* DOR Bt-1 Isolate," *Pest Technology* 6, (2012a): 79-82.

Reddy V P, Narasimha Rao N, Vimala Devi P S, Sivaramakrishnan S, Lakshmi Narasu M, Dinesh Kumar V, "Cloning, Characterization, and Expression of a New cry1Ab Gene from DOR Bt-1, an Indigenous Isolate of *Bacillus thuringiensis*," *Molecular Biotechnology*, (2012b), https://doi.org/10.1007/s12033-012-9627-3.

Reddy K R K, Rao G P, and Bahadur B, "*In vitro* Morphogenesis from Seedling Explants and Callus Cultures of Castor (*Ricinus communis* L.)," *Phytomorphology* 37, (1987): 337-340.

Sailaja M, Tarakeswari M, and Sujatha M, "Stable Genetic Transformation of Castor (*Ricinus communis* L.) *via* Particle Gun-mediated Gene Transfer using Embryo Axes from Mature Seeds," *Plant Cell Reports* 27, (2008): 509-1519.

Sai Sudha P, Tarakeswari M, Padmavathi A V T, Kumar O A, Chakravartty N, Vineeth K V, Mohan Katta A V S K, Sivarama P L, Boney K, Gupta S, Sujatha M,· and Lachagari V B R, "Deciphering the Transcriptomic Insight during Organogenesis in Castor (*Ricinus communis* L.), Jatropha (*Jatropha curcas* L.) and Sunflower (*Helianthus annuus* L.), *Biotech* 9, no. 434 (2019). https://doi.org/10.1007/s13205-019-1960-9.

Sarvesh A, Ram Rao D M, and Reddy T P, "Callus Initiation and Plantlet Regeneration from Epicotyl and Cotyledonary Explants of Castor (*Ricinus communis* L.)," *Advances in Plant Sciences* 5, (1992): 124-128.

Sharma H C, Sharma K K, Seetharama N, and Ortiz R, "Prospects for Using Transgenic Resistance to Insects in Crop Improvement," *Molecular Biology and Genetics* 3, no. 2, (2000): 1-28.

Shrirame H Y, Panwar N L, and Bamniya B, "Bio Diesel from Castor Oil-A Green Energy Option," *Low Carbon Economy* 2, no.1 (2011), https://doi.org/10.4236/lce.2011.21001.

Sousa N L, Cabral G B, Vieira P M, Baldoni A B, Francisco J L, "Bio-Detoxification of Ricin in Castor Bean (*Ricinus communis* L.) Seeds," *Scientific Reports* 7, (2017): 15385.

Singh P, Kumar M, Chaturvedi C P, Yadav D, and Tuli R, Development of a Hybrid Delta-endotoxin and its Expression in Tobacco and Cotton for Control of a Polyphagous Pest *Spodopteralitura" Transgenic Research* 13, (2004): 397-410.

Subbanna A R N S, Khan M S, Stanley J and KalyanaBabu B, "Diversity of *Bacillus thuringiensis* Isolates Native to Uttarakhand Himalayas, India and their Bio-efficacy against Selected Insect Pests," National Academy of Sciences, India, Section B: Biological Sciences, (2017), https://doi.org/10.1007/s40011-017-0892-6.

Sujatha M and Lakshminarayana M, "Susceptibility of Castor Semilooper, *Achaea janata* L. To Insecticide, Crystal Proteins from *Bacillus thuringiensis*," *Indian Journal of Plant Protection* 33, no. 2 (2005): 286-287.

Sujatha M and Reddy T P, "Differential Cytokinin Effects on the Stimulation of *in vitro* Shoot Proliferation from Meristematic Explants of Castor (*Ricinus communis* L.)," *Plant Cell Reports* 17, (1998): 561-566.

Sujatha M and Reddy T P, "Promotive Effect of Lysine Monohydrochloride on Morphogenesis in Cultured Seedling and Mature Plant Tissues of Castor (*Ricinus communis* L.)," *Indian Journal of Crop Science* 2, (2007): 279-286.

Sujatha M and Sailaja M, "Stable Genetic Transformation of Castor (*Ricinus communis* L.) via *Agrobacterium tumefaciens*-mediated Gene Transfer using Embryo axes from Mature Seeds," *Plant Cell Reports* 23, (2005): 803-810.

Sujatha M, Lakshminarayana M, Tarakeswari M, Singh P K, and Tuli R, "Expression of the cry1EC gene in Castor (*Ricinus communis* L.) Confers Field Resistance to Tobacco Caterpillar (*Spodoptera litura* Fabr) and Castor Semilooper (*Achaea janata* L.)," *Plant Cell Reports* 28, (2009): 935-946.

Sujatha M, Reddy T P, and Mahasi M J, "Role of Biotechnological Interventions in the Improvement of Castor (*Ricinus communis* L.) and *Jatropha curcas* L.," *Biotechnology Advances*, 26, (2008): 424-435.

Tarakeswari M, Ananda Kumar P, Lakshminarayana M, and Sujatha M, "Development and Evaluation of Transgenic Castor (*Ricinus communis* L.) Expressing the Insecticidal Protein Cry1Aa of *Bacillus thuringiensis* against Lepidopteran Insect Pests," *Crop Protection* 119, (2019): 113-125.

Travers R S, Martin P A W and Reichelderfer C F, "Selective Process for Efficient Isolation of Soil *Bacillus* spp.," *Applied and Environmental Microbiology* 53, no. 6 (1987): 1263-1266.

Varaprasad K S and Duraimurugan P, "Post-harvest Technology of Oilseeds," *Udyogprerana* 11, no. 2, (2016): 40-45.

Vimala Devi P S and Rao M L N, "Lab to Land Transfer Through Participatory Approach – the Case of *Bacillus thuringiensis* Within the Reach of the Dryland Farmer," *Journal of Rural Development* 24, no. 3 (2005a): 377-391.

Vimala Devi P S and Rao M L N, "Tailoring Production Technology: *Bacillus thuringiensis* (Bt) for Localized Production," *Tailoring Biotechnology* 1, no. 2 (2005b): 107-120.

Vimala Devi P S and Sudhakar R, "Effectiveness of a Local Strain of *Bacillus thuringiensis* in the Management of Castor Semilooper *Achaea janata* on Castor (*Ricinus communis*)," *Indian Journal of Agricultural Science* 76, no. 7 (2006): 447-449.

Vimala Devi P S and Vineela V, 2016 "Selection, Characterization and Potency Determination of a *Bacillus thuringiensis kurstaki* Isolate Toxic to Major Lepidopteran Pests," *Biopesticides International* 12, no.2 (2016): 139-148.

Vimala Devi P S, Balakrishnan K, Ravinder T and Prasad Y G, "Identification of Potent Strains of *Bacillus thuringiensis* for the Management of Castor Semilooper *Achaea janata* (Linn) and Optimization of Production," *Entomon,* 26, (2001): 98-103.

Vimala Devi P S, Duraimurugan P, Poorna Chandrika K S V, Vineela V, and Hari P P, "Novel Formulations of *Bacillus thuringiensis* var.

kurstaki: an Eco-friendly Approach for Management of Lepidopteran Pests," *World Journal of Microbiology and Biotechnology* 36, no.78 (2020), https://doi.org/10.1007/s11274-020-02849-8.

Vimala Devi P S, Duraimurugan P, Chandrika K S V P, and Vineela V, "Development of a Water Dispersible Granule (WDG) Formulation of *Bacillus thuringiensis* for the Management of *Spodopteralitura* (Fab.)," *Biocontrol Science and Technology*, (2021), https://doi.org/ 10.1080/ 09583157.2021.1895073.

Vimala Devi P S, Ravinder T, and Jaidev C, "Cost-effective Production of *Bacillus thuringiensis* by Solid-State Fermentation," *Journal of Invertebrate Pathology*88, no. 2 (2005): 163-168.

In: *Bacillus thuringiensis*
Editor: David P. Sanders

ISBN: 978-1-53619-570-5
© 2021 Nova Science Publishers, Inc.

Chapter 3

STRATEGIES FOR BOOSTING OR MAINTAINING THE INSECTICIDAL ACTIVITY OF *BACILLUS THURINGIENSIS*

Wataru Mitsuhashi

Institute of Agrobiological Sciences,
National Agriculture and Food Research Organization (NARO)
Tsukuba, Ibaraki, Japan

ABSTRACT

Cry toxins, one category of δ-endotoxins that are produced by the bacterium *Bacillus thuringiensis* (Bt), are used in sprays and genetically transformed crops to control insect pests. However, the rapid evolution of Bt Cry toxin resistance in target pests threatens the potency of these proteins. Thus, there is a need to elucidate the mechanism of Bt resistance and to develop strategies to cope with such resistance. Furthermore, strengthening the efficacy of Cry toxins against Bt-susceptible pests is also important. In this chapter, we first present a brief overview of δ-endotoxins. Next, we review relevant studies and discuss strategies to mitigate the spread of Bt resistance and improve insecticidal activity against Bt-susceptible pests.

Keywords: *Bacillus thuringiensis, Bombyx mori,* 3D-Cry toxins, Bt resistance, peritrophic matrix, enhanced insecticidal activity

INTRODUCTION

The entomopathogenic bacterium *Bacillus thuringiensis* (Bt) produces several types of insecticidal proteins: δ-endotoxins, which are formed as crystals in bacterial cells during the sporulation phase, and vegetative insecticidal protein (Vip) toxins and secreted insecticidal protein (Sip) (Donovan et al., 2006), both of which are secreted from cells during the growth phase and do not form crystals. Deruta-endotoxins are categorized into two types, Cry and Cyt toxins.

The Bt Toxin Nomenclature Committee has classified Cry toxins into 73 different types (Cry1 to Cry73) (Palma et al., 2014). Most Cry toxins are produced by Bt, but other toxins such as Cry16 and Cry17, Cry18, Cry43, and Cry48 and Cry49 are produced by *Clostridium bifermentans, Paenibacillus popilliae, Paenibacillus lentimorbus,* and *Lysinibacillus sphaericus,* respectively. Three domain Cry toxins (3D-Cry toxins) are the largest category of Cry toxins and consist of domains I, II, and III. There are also other types of Cry toxins, such as Bin-like toxins, Mtx-like toxins, and parasporins (Palma et al., 2014; Sato, 2014).

Cyt toxins have a different structure from Cry toxins, a small 25-30 kDa size (Hayakawa and Sakai, 2008), and are broadly toxic to both mammals and insects (Crickmore et al., 1998).

The 3D-Cry toxins have a restricted spectrum for killing insect species and are not toxic to human and animals. Therefore, 3D-Cry toxins have been widely used as microbial insecticides to control insect pests and expressed within crops as Bt crops to control pests. However, the evolution of Bt resistance (resistance against 3D-Cry toxins) in insects has threatened pest control using 3D-Cry toxins. Therefore, better understanding of the mechanism of Bt resistance and developing strategies to overcome Bt resistance are required, which is closely related to further progress in the elucidation of the Bt Cry toxin mode of action. Improving the insecticidal activity of toxins to which the evolution of resistance of insects has not been reported is also important. These new technologies and strategies will lead to the global spread of the usage of Bt (Figure 1).

In this review, we first present a brief overview of insecticidal proteins produced by Bt and the mode of action of Cry toxins. Next, we review studies and strategies to enhance the potency of Cry toxins to prevent the spread of Bt resistance and improve the insecticidal activity of Cry toxins against Bt-susceptible pests.

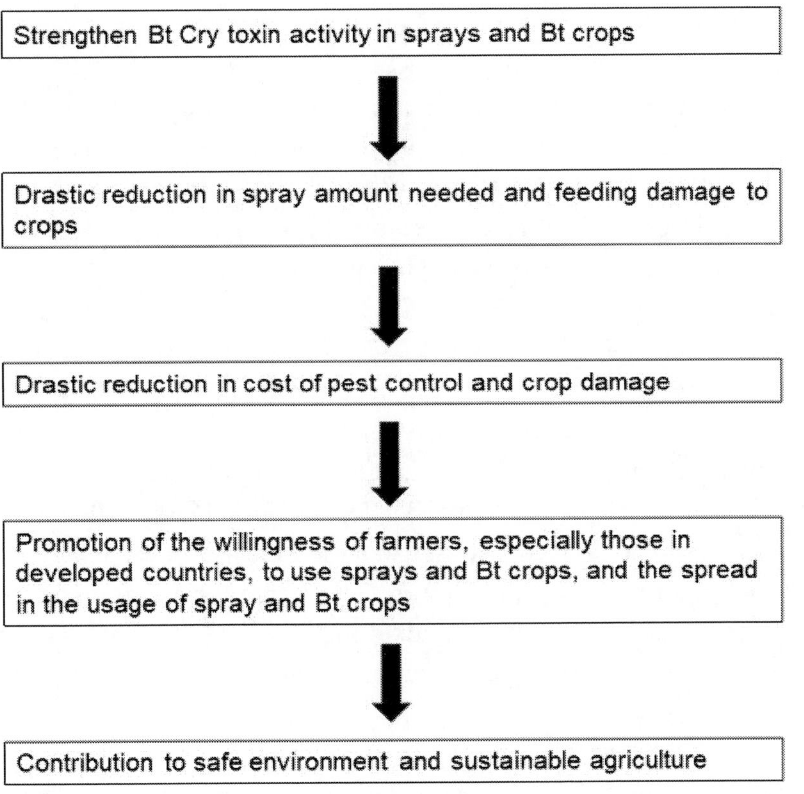

Figure 1. Pathway for the realization of the widespread use of Bt technologies in agriculture.

δ-ENDOTOXINS

3D-Cry Toxins

Information on the toxin structures and spectra in this section is based on Palma et al. (2014) and Sato (2014).

Most 3D-Cry toxins are produced as 120-130 kDa prototoxins except for the prototoxins of Cry2, Cry3, Cry10, Cry11, Cry13, and Cry1I, which have size in the range 70-80 kDa. Many 3D-Cry toxin genes are present in large plasmids. In many cases, one plasmid harbors multiple kinds of 3D-Cry toxin genes.

Domain I is a perforating domain, consisting of a cluster of seven α-helixes, and is associated with toxin membrane insertion and pore-formation. Domain II is a beta-prism I, involved in toxin-receptor interactions. Domain III is a galactose-binding domain and has a β-sandwich structure. Domain III is involved in receptor binding and pore formation.

Cry4, Cry10, Cry11, Cry16, Cry17, Cry19, Cry20, Cry24, Cry27, Cry29, Cry30, Cry32, Cry39, Cry40, Cry44, Cry47, Cry48, and Cry52 are toxic to Diptera. Cry3, Cry7, Cry8, Cry18, and Cry43 are toxic to Coleoptera, and Cry5, Cry12, Cry13, Cry14, and Cry21 are toxic to Nematoda. Cry1 is mainly toxic to Lepidoptera, but Cry1Ba is also toxic to Diptera, leaf beetles, and bark beetle, and Cry1Ca is also toxic to mosquitoes. In addition, Cry1Ia is also toxic to leaf beetles, and Cry1Ac is toxic to tsetse flies and aphids. Cry2, Cry54, and Cry56 are toxic to both Lepidoptera and Diptera. Cry9 is toxic to Lepidoptera. Mode of action of 3D-Cry toxins is described in the section below.

Cry Toxins Other than 3D-Cry Toxins

Cry toxins include Mtx-like toxins, Bin-like toxins, Cry6, 22, 37 group, parasporin, and others. Toxin features (Sato, 2014) are described below, but details are shown in Palma et al. (2014).

Mtx-like proteins are approximately 40 kDa in size, and Cry15, Cry23, Cry33, and Cry38 belong to this category. Mtx-like proteins are named because their amino acid sequences are similar to those of the mosquitocidal toxins, Mtx2 and Mtx3, which are produced by *L. sphaericus*. Mtx-like proteins are also similar to ε toxin in *C. perfringens*, which is a pore-forming toxin. Therefore, the mode of action of Mtx-like proteins may be similar to that of 3D-Cry toxins (Popoff, 2011). Cry15 is toxic to Lepidoptera, and Cry23 and Cry38 are toxic to Coleoptera.

Bin-like proteins are approximately 44 kDa in size. Cry35, Cry36, and Cry49 are included in this category. Bin-like proteins are named because of their structural similarity to binary BinA/BinB toxin from *L. sphaericus*. These proteins are estimated to be beta pore-forming toxins from the crystal structure of Cry35 (Kelker et al., 2014). Binary toxins Cry34/Cry35 and Cry48/Cry49 are toxic to Coleoptera and Diptera, respectively. Cry36 is toxic to Coleoptera.

Cry34 is 14 kDa, and its amino acid sequence is similar to the hemolysin aerolysin that forms pores in membranes.

Cry6, Cry22, and Cry37 are placed in a group because of amino acid homology, and Cry6, Cry22, and Cry37 proteins have 475, 722, and 126 amino acid residues, respectively. Cry6, Cry22, and Cry37 are toxic to Nematoda, Chrysomelidae, and weevils; ants; and Scarabaeidae, respectively.

Parasporins are defined as *Bt* and related bacterial parasporal proteins that are non-hemolytic but are able to preferentially kill cancer cells. However, the toxicity of parasporins in insects remains unknown (Chubicka et al., 2018). Parasporins are classified into six groups of PS1 to PS6 based on their structure and activity level (Okumura, 2010). PS1 (Cry31) has a 3-D toxin structure but kills cancer cells through apoptosis. PS2 is structurally a β-pore-forming toxin. The structure of PS3 (Cry41) has a C terminus similar to that of a 3D-Cry toxin with a ricin B-like lectin domain. PS4 (Cry45) is a β-pore-forming toxin that is similar to an Mtx-like toxin that kills cancer cells by forming pores.

Cyt Toxins

Cyt toxins mainly exhibit toxicity against mosquitos, but they also have hemolytic activity against mammal cells and hemocytes. They are classified into Cyt1, Cyt2, and Cyt3 families. The atomic structure of Cyt2Aa reveals that it is a one-domain protein and is similar to Volvatoxin from the mushroom *Volvariella volvacea*. Proteins in the Cyt families have a small 25–30 kDa size (Hayakawa and Sakai, 2008) and are not phylogenetically related to 3D-Cry toxins. Two modes of action have been proposed: one is a pore-formation model and another supports a less specific detergent action mechanism (Butko, 2003; De Maagd et al., 2003; Cohen et al., 2011; Soberón et al., 2013).

VIP TOXINS

Vips are secreted from Bt during the growth phase and are classified into Vip1, Vip2, Vip3, and Vip4. There is no structural similarity among Vip1, Vip2, and Vip3 proteins. Vip1, Vip2, Vip3, and Vip4 are approximately 100 kDa, 53 kDa, 90 kDa, and 108 kDa, respectively. The amino acid sequence of Vip1 and Vip2 is similar to that of a protective antigen in *B. anthracis* and C2 toxin in *Clostridium botulinum*, respectively. Vip3 contains a domain of the carbohydrate binding module 4/9 family. The Vip1 and Vip2 toxins constitute a binary toxin, and the binary Vip1/Vip2 toxin are toxic to Lepidoptera. Vip1 oligomers form pores in midgut cell membranes and contribute the entry of Vip2 into the cells. Vip3 also forms pores in midgut cell membranes after activation by digestive enzymes, but the mode of action remains poorly understood (Palma et al., 2014). Vip3 are also toxic to Lepidoptera. Insecticidal properties of Vip4 are unknown.

SIP TOXIN

Sip1Aa1 (Donovan et al., 2006) shows low similarity to the 36-kDa Mtx3 mosquitocidal toxin from *L. sphaericus*. This suggests that the mode of action of Sip1Aa1 may be pore formation, but this remains unknown (Palma et al., 2014). It is approximately 40 kDa in size and toxic to Coleoptera.

MODE OF ACTION OF CRY TOXINS

3D-Cry toxins are the most-extensively studied group among Bt toxins. The mode of action of 3D-toxins has been investigated intensively. The N and C termini of 3D-Cry protoxins fed to insects are cleaved by midgut proteases, which activates the toxin. 3D-Cry proteins then pass through the peritrophic matrix (PM), which is a noncellular, tubular membrane that lines the midgut lumen, extending from the anterior midgut to the hindgut, and is composed of chitin fibers, proteins, glycoproteins, and proteoglycans (Derksen and Granados, 1988; Hegedus et al., 2019). The toxins specifically lyse the midgut cells to disrupt the midgut epithelium, leading to septicemia and eventually mortality in insects (Raymond et al., 2010; Schnepf et al., 1998; Heckel, 2016; Wang et al., 2018). Models of the toxic action of Cry1A toxins include midgut proteins such as a cadherin, an ABC transporter, a membrane bound alkaline phosphatase, and an aminopeptidase N as receptors for the toxins, and the receptors are proposed to function sequentially with the toxins to form pores in the cell membrane (Bretschneider et al., 2016; Pardo-Lopez et al., 2013; Wang et al., 2018) or to function as a receptor in an adenylyl cyclase/PKA signaling pathway to induce necrotic cell death (Zhang et al., 2005). Both the sequential receptor binding model and the adenylyl cyclase/PKA signaling pathway model suggest a single pathway of action for Cry1A toxins (Wang et al., 2018).

Wang et al. (2018), however, reported that cadherin and ABCC2, a type of ABC transporter, independently play a role in the mode of action of Cry1A toxins and thus the toxins exert insecticidal activity through multiple redundant pathways of toxicity in insects.

STRATEGIES FOR ENHANCING OR MAINTAINING THE INSECTICIDAL ACTIVITY OF BT

Settlement of Non-Bt Refuge

Cultivation of wild-type crops near Bt-resistant crops has been used to prevent the evolution of Bt-resistant pests (Gould, 1998; Carrie `Re et al., 2001; Campagne et al., 2016). In settlement of non-Bt refuges, Bt crops should produce very high doses of toxin(s) relative to LD_{99} of target insects (Gould, 1998), and the refuge size should be 5% of area of Bt cultivation without pest control or 20% of that when pest control is performed with insecticides other than Bt insecticide (Gould, 1998).

In addition, a reversal of resistance occurs when Bt exposure was stopped for many generations (Tabashnik et al., 1994). Therefore, one mitigation strategy is to stop cultivation of GM crops until Bt-resistant pests diminish.

Use of Additives in Bt-Formulations

The PM protects the luminal surface from invasion by pathogenic microorganisms (Hegedus et al., 2009). In addition, Cry toxins bind to or are trapped by the PM (Rees et al., 2009; Denolf et al., 1993; Hayakawa et al., 2004). The level of blocking of Cry toxins by the PM depends on the kinds of the toxin and insect species (Hayakawa et al., 2004; Bravo et al., 1992; Bravo et al., 1992). The baculoviral protein enhancin, the entomopoxviral and baculoviral fusolins, and some chemicals are known to disrupt the PM (Chiu et al., 2015; Derksen and Granados, 1988; Mitsuhashi et al., 1998; Wang and Granados, 2001; Mukawa et al., 2003; Mitsuhashi et al., 2007; Takemoto et al., 2008; Liu et al., 2011; Mitsuhashi and Miyamoto, 2003). Thus, these materials increase the insecticidal activity of Bt and Bt Cry toxins in several insects, as described below.

terminal region exposed a hidden receptor-binding region that improved toxicity (Morse et al., 2001; Mandal et al., 2007; Tabashnik et al., 2013).

Soberón et al. (2007) found that modified Cry1Ab and Cry1Ac lacking helix α-1 formed oligomers *in vitro* without cadherin and killed Cry1A-resistant insects, which suggested that oligomer formation without receptor binding can counter Bt resistance. Another strategy is to modify Cry toxin to facilitate passage of toxins through the PM.

Mixtures of Bt Toxins

Synergistic effects of mosquitocidal activity were observed when Cry4A was mixed with both Cry4B and Cry4D or Cry4B was mixed with Cry4D (Poncet et al., 1995). These toxins are components of *B. thuringiensis* subsp. *israelensis* toxin, and resistance to the toxin of *israelensis* has not been reported in field populations (Ben-Dov, 2014). These results suggested synergistic interactions among these toxins suppress the evolution of resistance. The lack of resistance to the toxin may be attributed partly to the use of different receptors by the components. Cry46Ab, which was deduced to be a member of the aerolysin-type B pore-forming toxins, enhanced the mosquitocidal activity of 3D-Cry toxins such as Cry4Aa, Cry11Aa, and Cry11Ba (Ito et al., 2004; Sher et al., 2005; Hayakawa et al., 2017). This also suggested that the co-use of Cry46Ab and 3D-Cry toxins may suppress the prevalence of mosquitos resistant to 3-D toxins.

Cyt1Aa toxin can enhance the activity of Cry4Aa, Cry4Ba, and Cry11Aa toxins against mosquitoes (Pérez et al., 2005; Cantón et al., 2011; Wirth et al., 1997). The binding of Cry11Aa to *Aedes aegypti* brush border membrane vesicles was enhanced by membrane-bound Cyt1Aa (Pérez et al., 2005). These results strongly suggested that co-use of 3D-and Cyt toxins can counter some aspects of Bt resistance in mosquitoes.

Modification of Traditional Transgenic Crop Usage

Stacked-trait Bt crops expressing multiple Cry toxins have been replacing single-transgene GM crops, and up to six Cry toxins can be expressed in a single crop. However, combining multiple Cry toxins raises the question of possible risks from toxin–toxin interactions, and no guidelines for assessing strategies and procedures have been authorized (Hilbeck and Otto, 2015).

Expression of the viral proteins enhancin and fusolin in GM crops enhanced the susceptibility of some pest insects to baculoviral insecticides compared with wild-type crops (Hayakawa et al., 2000; Hukuhara et al., 1999). Similarly, crops that express both Bt toxins and these viral proteins or viral proteins alone are likely to be more effective in killing certain insect pests.

Negative Cross-Resistance

Xiao et al. (2016) reported that resistance to Cry1Ac caused by a mutation in the ABCC2 transporter is coupled to increased susceptibility to bacterial insecticides, such as abamecin and spineotram, derived from the soil bacteria *Streptomyces avermitilis* and *Saccharopolyspora spinosa*, *respectively*. These results suggested these bacterial insecticides are potential candidates to reduce Cry1Ac resistance.

Searches for counterparts of Bt toxins that show negative cross-resistance will be promoted by using information of well-annotated insect genome sequences, including that of *B. mori*.

Other Means

Before some of these measures were implemented, Cry toxins, including Cry1Aa and some types of Cry9, would be used against lepidopteran pests because evolution of insect resistance to such toxins is not conspicuous (Shu et al., 2013).

Finding novel proteins is very important to counter field-evolved resistance. Next-generation sequencing technologies (Mardis, 2008; Zhou et al., 2010) are promising tools to aid the discovery of novel toxin genes.

Mutations at several sites in ABCC2 other than a mutation reported by Atsumi et al. (2012) do not induce death in *B. mori* larvae (Kazuhisa Miyamoto, personal communication), which suggests that this protein does not naturally play an important role in insect survival or that the loss of this protein is compensated by other protein functions. Thus, researchers may search for Cry toxins that bind to new types of receptors necessary for survival where Bt-resistant pests with mutations in these receptors cannot proliferate and die.

CONCLUSION

Main sources of microbial control are Bt Cry toxins, viruses, fungi, nematoda, and protozoa. Bt Cry toxins are better than the other sources of microbial control because of their faster control of insects and also cheaper production cost than viruses and protozoa. Therefore, they are one of the most potent agents for pest control, and further studies to expand their usage as insecticides and in Bt crops are widely expected. In many crop fields, however, it will be better for effectiveness of pest control if Bt toxins are combined with some chemical insecticides, and thus the elucidation of optimum combinations is an important issue.

A combination or integration of individual strategies may enhance the effectiveness of each countermeasure against Bt resistance, but these will vary depending on the type of resistance. Therefore, research into combination or integration of pest control strategies is important. In addition, Cry toxin agents or potential agents mainly target lepidopteran pests; therefore, searches for agents capable of targeting pests in other orders is required.

Improvement of genetically transformed crops also an important strategy for the sustainable usage of Bt. For this purpose, improved technology to prevent environmental pollution by Bt crop pollen is required (Hüsken et al., 2010) in addition to improvement of the expression levels of Cry genes.

In addition, the use of Bt is not restricted to pest control. Applications such as dual control of pests and plant infectious diseases and for cancer treatment with parasporins are being expanded (Takahashi et al., 2014). Bt may not originally be an insect bacterium because it does not proliferate in insects, and this aspect may lead to further uses in other fields in the future.

REFERENCES

Arakawa, T. (2003). Chitin synthesis inhibiting antifungal agents promote nucleopolyhedrovirus infection in silkworm, *Bombyx mori* (Lepidoptera: Bombycidae) larvae. *Journal of Invertebrate Pathology*, *83*, 261–263. doi: 10.1016/S0022-2011(03)00085-5.

Atsumi, S., Miyamoto, K., Yamamoto, K., Narukawa, J., Kawai, S., Sezutsu, H., Kobayashi, I., Uchino, K., Tamura, T., Mita, K., Kadono-Okuda, K., Wada, S., Kanda, K., Goldsmith, M. R. & Noda, H. (2012). Single amino acid mutation in an ATP-binding cassette transporter gene causes resistance to Bt toxin Cry1Ab in the silkworm, *Bombyx mori*. *Proceedings of the National Academy of Sciences of the United States of America*, *109*, E1591–E1598. doi: 10.1073/pnas.1120698109.

Bah, A., van Frankenhuyzen, K., Brousseau, R. & Masson, L. (2004). The *Bacillus thuringiensis* Cry1Aa toxin: Effects of trypsin and chymotrypsin site mutations on toxicity and stability. *Journal of Invertebrate. Pathology*, *85*, 120–127. doi: 10.1016/j.jip.2004.02.002.

Ben-Dov, E. (2014). *Bacillus thuringiensis* subsp. *israelensis* and its dipteran-specific toxins. *Toxins*, *6*, 1222–1243. doi: 10.3390/toxins6041222.

Berliner E. (1915). Über die Schlaffsucht der Mehlmottenraupe (*Ephestia kűhniella*, Zell) und ihren Erreger *Bacillus thuringiensis*, n. sp.

Zeitschrift fur Angewandte Entomologie, *2*, 29–56. doi: 10. 1111/j.1439-0418.1915.tb00334.x. [About the drowsiness of the flour moth caterpillar (*Ephestia kűhniella*, Zell) and its pathogen Bacillus thuringiensis, n. *Journal of Applied Entomology*]

Bravo, A., Jansens, S. & Peferoen, M. (1992). Immunocytochemical localization of *Bacillus thuringiensis* insecticidal crystal proteins in intoxicated insects. *Journal of Invertebrate Pathology*, *60*, 237–246.

Bravo, A., Hendrichx, K., Jansens, S. & Peferoen, M. (1992). Immunocytochemical analysis of specific binding of *Bacillus thuringiensis* insecticidal crystal proteins to lepidopteran and coleopteran midgut membranes. *Journal of Invertebrate Pathology*, *60*, 247–253.

Bretschneider, A., Heckel, D. G. & Pauchet, Y. (2016). Three toxins, two receptors, one mechanism: mode of action of Cry1A toxins from *Bacillus thuringiensis* in *Heliothis virescens*. *Insect Biochemistry and Molecular Biology*, *76*, 109–117. doi: 10.1016/j.ibmb.2016.07.008.

Butko, P. (2003). Cytolytic toxin Cyt1A and its mechanism of membrane damage: Data and hypotheses. *Applied and Environmental Microbiology*, *69*, 2415–2422. doi: 10.1128/AEM.69.5.2415-2422.2003.

Cantón, P. E., Reyes, E. Z., RuízdeEscudero, I., Bravo, A. & Soberón, M. (2011). Binding of *Bacillus thuringiensis* subsp. *israelensis* Cry4Ba to Cyt1Aa has an important role in synergism. *Peptides*, *32*, 595–600. doi: 10.1016/j.peptides.2010.06.005.

Carrie `Re, Y., Dennehy, T. J., Pedersen, B., Haller, S., Ellers-Kirk, C., Antilla, L., Liu, Y. B., Willott, E. & Tabashnik, B. E. (2001). Large-scale management of insect resistance to transgenic cotton in Arizona: Can transgenic insecticidal crops be sustained? *Journal of Economic Entomology*, *94*, 315–325.

Campagne, P., Smouse, P. E., Pasquet, R., Silvain, J. F., Le Ru, B. & Van den Berg, J. (2016). Impact of violated high-dose refuge assumptions on evolution of Bt resistance. *Evolutionary Applications*, *9*, 596–607. doi: 10.1111/eva.12355.

Chiu, E., Hijnen, M., Bunker, R., Boudes, M., Rajendran, C., Aizel, K., Olieric, V., Schulze-Briese, C., Mitsuhashi, W., Young, V., Ward, V. K., Bergoin, M., Metcalf, P. & Coulibaly, F. (2015). Structural basis for the enhancement of virulence by viral spindles and their *in vivo* crystallization. *Proceedings of the National Academy of Sciences of the United States of America*, *112*, 3973–3978. doi: 10.1073/pnas.1418798112.

Chubicka, T., Girija, D., Deepa, K., Salini, S., Meera, N., Raghavamenon, A. C., Divya, M. K. & Babu, T. D. (2018). A parasporin from *Bacillus thuringiensis* native to Peninsular India induces apoptosis in cancer cells through intrinsic pathway. *Journal of Biosciences*, *43*, 407–416.

Cohen, S., Albeck, S., Ben-Dov, E., Cahan, R., Firer, M., Zaritsky, A. & Dym, O. (2011). Cyt1Aa toxin: Crystal structure reveals implications for its membrane-perforating function. *Journal of Molecular Biology*, *413*, 804–814. doi: 10.1016/j.jmb.2011.09.021.

Dean, M. & Annilo, T. (2005). Evolution of the ATP-binding cassette (ABC) transporter superfamily in vertebrates. *Annual Review of Genomics and Human Genetics*, *6*, 123–142. doi: 10.1146/annurev.genom.6.080604.162122.

Deist, B. R., Rausch, M. A., Fernandez-Luna, M. T., Adang, M. J. & Bonning, B. C. (2014). Bt toxin modification for enhanced efficacy. *Toxins*, *6*, 3005–3027. doi: 10.3390/toxins6103005.

Denolf, P., Jansens, S., Peferoen, M., Degheels, D. & van Rie, J. (1993). Two different *Bacillus thuringiensis* delta-endotoxin receptors in the midgut brush border membrane of the European corn borer, *Ostrinia nubilalis* (Hiubner) (Lepidoptera: Pyralidae). *Applied and Environmental Microbiology*, *59*, 1828–1837.

Derksen, A. C. G. & Granados, R. R. (1988). Alteration of a lepidopteran peritrophic membrane by baculoviruses and enhancement of viral infectivity. *Virology*, *167*, 242–250. doi: 10.1016/0042-6822(88)90074-8.

Donovan, W. P., Engleman, J. T., Donovan, J. C., Baum, J. A., Bunkers, G. J., Chi, D. J., Clinton, W. P., English, L., Heck, G. R., Ilagan, O. M., Krasomil-Osterfeld, K. C., Pitkin, J. W., Roberts, J. K. & Walters, M.

R. (2006). Discovery and characterization of Sip1A: A novel secreted protein from *Bacillus thuringiensis* with activity against coleopteran larvae. *Applied Microbiology and Biotechnology, 72,* 713–719. doi: 10.1007/s00253-006-0332-7.

Gould, F. (1998). Sustainability of transgenic insecticidal cultivars: Integrating pest genetics and ecology. *Annual Review of Entomology, 43,* 701–26. doi: 10.1146/annurev.ento.43.1.701.

Granados, R. R., Fu, Y., Corsaro, B. & Hughes, P. R. (2001). Enhancement of *Bacillus thuringiensis* toxicity to lepidopterous species with the enhancin from *Trichoplusia ni* granulovirus. *Biological Control, 20,* 153–159. doi: 10.1006/bcon.2000.0891.

Greiner, T., Moroni, A., Etten, J. L. V. & Thiel, G. (2018). Genes for membrane transport proteins: not so rare in viruses. *Viruses, 10,* 456. doi: 10.3390/v10090456.

Hayakawa, T. & Sakai, Y. (2008). Possibility of applied use of Cry toxins with special reference to parasporins. *Sanshi-Kontyu Biotec, 77,* 205–210. (in Japanese).

Hayakawa, T., Sakakibara, A., Ueda, S., Azuma, Y., Ide, T. & Takebe, S. (2017). Cy46Ab from *Bacillus thuringiensis* TK-E6 is a new mosquitocidal toxin with aerolysin-type architecture. *Insect Biochemistry and Molecular Biology, 87,* 100–106. doi: 10.1016/j.ibmb.2017.06.015.

Hayakawa, T., Shitomi, Y., Miyamoto, K. & Hori, H. (2004). GalNAc pretreatment inhibits trapping of *Bacillus thuringiensis* Cry1Ac on the peritrophic membrane of *Bombyx mori. FEBS Letters, 576,* 331–335. doi: 10.1016/j.febslet.2004.09.029.

Hayakawa, T., Shimojo, E., Mori, M., Kaido, M., Furusawa, I., Miyata, S., Sano, Y., Matsumoto, T., Hashimoto, Y. & Granados, R. R. (2000). Enhancement of baculovirus infection in *Spodoptera exigua* (Lepidoptera: Noctuidae) larvae with *Autographa californica* nucleopolyhedrovirus or *Nicotiana tabacum* engineered with a granulovirus enhancin gene. *Applied Entomology and Zooogy, 35,* 163–170. doi: 10.1303/aez.2000.163.

Heckel, D. G. (2012). Learning the ABCs of Bt: ABC transporters and insect resistance to *Bacillus thuringiensis* provide clues to a crucial step in toxin mode of action. *Pesticide Biochemistry and Physiology, 104,* 103–110. doi: 10.1016/j.pestbp.2012.05.007.

Heckel, D. G. (2016). A return to the pore - dissecting *Bacillus thuringiensis* toxin mode of action via voltage clamp experiments. *FEBS Journal, 283,* 4458–4461. doi: 10.1111/febs.13973.

Hegedus, D., Erlandson, M., Gillott, C. & Toprak, U. (2009). New insight into peritrophic matrix synthesis, architecture, and function. *Annual Review of Entomology, 54,* 285–302. doi: 10.1146/annurev.ento.54.110807.090559.

Hegedus, D. D., Toprak, U. & Erlandson, M. (2019). Peritrophic matrix formation. *Journal of Insect Physiology, 117,* 103898. doi: 10.1016/j.jinsphys.2019.103898.

Hilbeck, A. & Otto, M. (2015). Specificity and combinatorial effects of *Bacillus Thuringiensis* Cry toxins in the context of GMO environmental risk assessment. *Frontiers in Environmental Science, 3,* Article 71. doi: 10.3389/fenvs.2015.00071.

Himeno, M. (1999). Improvement and mechanism of action of microbial pesticides (Bt). *Microbes* and *Environments, 14,* 245–252. (in Japanese). doi: 10.1264/jsme2.14.245.

Hukuhara, T., Hayakawa, T. & Wijonarko, A. (1999). Increased baculovirus susceptibility of armyworm larvae feeding on transgenic rice plants expressing an entomopoxvirus gene. *Nature Biotechnology, 17,* 1122–1124. doi: 10.1038/15110.

Hüsken, A., Prescher, S. & Schiemann, J. (2010). Evaluating biological containment strategies for pollen-mediated gene flow. *Environmental Biosafety Research, 9,* 67–73. doi: 10.1051/ebr/2010009.

Ito, A., Sasaguri, Y., Kitada, S., Kusaka, Y., Kuwano, K., Masutomi, K., Mizuki, E., Akao, T. & Ohba, M. (2004). A *Bacillus thuringiensis* crystal protein with selective cytocidal action to human cells. *Journal of Biological Chemistry, 279,* 21282–21286. doi: 10.1074/jbc. M401881200.

Kelker, M. S., Berry, C., Evans, S. L., Pai, R., McCaskill, D. G., Wang, N. X., Russell, J. C., Baker, M. D., Yang, C., Pflugrath, J. W., Wade, M., Wess, T. J. & Narva, K. E. (2014). Structural and biophysical characterization of *Bacillus thuringiensis* insecticidal proteins Cry34Ab1 and Cry35Ab1. *PLoS One, 9,* e112555. doi: 10.1371/journal.pone.0112555.

Konno, K. & Mitsuhashi, W. (2019). The peritrophic membrane as a target of proteins that play important roles in plant defense and microbial attack. *Journal of Insect Physiology, 117,* 10391. doi: 10.1016/j.jinsphys.2019.103912.

Liu, X., Ma, X., Lei, C., Xiao, Y., Zhang, Z. & Sun, X. (2011). Synergistic effects of *Cydia pomonella* granulovirus GP37 on the infectivity of nucleopolyhedro viruses and the lethality of *Bacillus thuringiensis. Archives of Virology, 156,* 1707–1715. doi: 10.1007/s00705-011-1039-3.

De Maagd, R. A., Bravo, A., Berry, C., Crickmore, N. & Schnepf, H. E. (2003). Structure, diversity, and evolution of protein toxins from spore-forming entomopathogenic bacteria. *Annual Review of Genetics, 37,* 409–433. doi: 10.1146/annurev.genet.37.110801.143042.

Mandal, C. C., Gayen, S., Basu, A., Ghosh, K. S., Dasgupta, S., Maiti, M. K. & Sen, S. K. (2007). Prediction-based protein engineering of domain I of Cry2A entomocidal toxin of *Bacillus thuringiensis* for the enhancement of toxicity against lepidopteran insects. *Protein Engineering, Design* and *Selection, 20,* 599–606. doi: 10.1093/protein/gzm058.

Mardis, E. R. (2008). Next-generation DNA sequencing methods. *Annual Review of Genomics and Human Genetics, 9,* 387–402. doi: 10.1146/annurev.genom.9.081307.164359.

Mitsuhashi, W., Asano, S., Miyamoto, K. & Wada, S. (2014). Further research on the biological function of inclusion bodies of *Anomala cuprea* entomopoxvirus, with special reference to effect on the insecticidal activity of a *Bacillus thuringiensis* formulation. *Pest Management Science, 70,* 46–54. doi: 10.1002/ps.3521.

Mitsuhashi, W., Furuta, Y. & Sato, M. (1988). The spindles of an entomopoxvirus of *Coleoptera* (*Anomala cuprea*) strongly enhance the infectivity of a nucleopolyhedrovirus in *Lepidoptera* (*Bombyx mori*). *Journal of Invertebrate Pathology, 71*, 186–188.

Mitsuhashi, W., Kawakita, H., Murakami, R., Takemoto, Y., Saiki, T., Miyamoto, K. & Wada, S. (2007). Spindles of an entomopoxvirus facilitate its infection of the host insect by disrupting the peritrophic membrane. *Journal of Virology, 81*, 4235–4243. doi: 10.1128/JVI.02300-06.

Mitsuhashi, W. & Miyamoto, K. (2020). Interaction of *Bacillus thuringiensis* Cry toxins and the insect midgut with a focus on the silkworm (*Bombyx mori*) midgut. *Biocontrol Science and Technology, 30*, 6884. doi: 10.1080/09583157.2019.1684439.

Morse, R. J., Yamamoto, T. & Stroud, R. M. (2001). Structure of Cry2Aa suggests an unexpected receptor binding epitope. *Structure, 9*, 409–417. doi: 10.1016/s0969-2126(01)00601-3.

Mukawa, S., Nakai, M., Okuno, S., Takatsuka, J. & Kunimi, Y. (2003). Nucleopolyhedrovirus enhancement by a fluorescent brightener in *Mythimna separata* (Lepidoptera: Noctuidae). *Applied Entomology and Zoology, 38*, 87–96. doi: 10.1303/aez.2003.87.

Okumura, S., Ohba, M., Mizuki, E., Crickmore, N., Cote, J. C., Nagamatsu, Y., Kitada, S., Sakai, H., Harata, K. & Shin, T. (2010). *Parasporin nomenclature*. (http://parasporin.fitc.pref.fukuoka.jp/).

Palma, L., Muñoz, D., Berry, C., Murillo, J. & Caballero, P. (2014). *Bacillus thuringiensis* toxins: an overview of their biocidal activity. *Toxins, 6*, 3296–3325. doi: 10.3390/toxins6123296.

Pardo-Lopez, L., Soberon, M. & Bravo, A. (2013). *Bacillus thuringiensis* insecticidal three domain Cry toxins: mode of action, insect resistance and consequences for crop protection. *FEMS Microbiology Reviews, 37*, 3–22. doi: 10.1111/j.1574-6976.2012.00341.x.

Pérez, C., Fernandez, L. E., Sun, J., Folch, J. L., Gill, S. S., Soberón, M. & Bravo, A. (2005). *Bacillus thuringiensis* subsp. Israelensis Cyt1Aa synergizes Cry11Aa toxin by functioning as a membrane-bound receptor. *Proceedings of the National Academy of Sciences of the United States of America, 102,* 18303–18308. doi: 10.1073/pnas.0505494102.

Popoff, M. R. (2011). Epsilon toxin: A fascinating pore-forming toxin. *FEBS Journal, 278,* 4602–4615. doi: 10.1111/j.1742-4658.2011.08145.x.

Poncet, S., Delécluse, A., Klier, A. & Rapoport, G. (1995). Evaluation of synergistic interactions among the CryIVA, CryIVB, and CryIVD toxic components of *B. thuringiensis* subsp. israelensis crystals. *Journal of Invertebrate Pathology, 66,* 131–135. doi: 10.1006/jipa.1995.1075.

Raymond, B., Johnston, P. R., Nielsen-LeRoux, C., Lereclus, D. & Crickmore, N. (2010). *Bacillus thuringiensis*: an impotent pathogen? *Trends in Microbiology, 18,* 189–194. doi: 10.1016/j.tim.2010.02.006.

Rees, J. S., Jarrett, P. & Ellar, D. J. (2009). Peritrophic membrane contribution to Bt Cry δ-endotoxin susceptibility in Lepidoptera and the effect of Calcofluor. *Journal of Invertebrate Pathology, 100,* 139–146. doi: 10.1016/j.jip.2009.01.002.

Saraswathy, N. & Kumar, P. A. (2004). Protein engineering of δ-endotoxins of *Bacillus thuringiensis*. *Electronic Journal of Biotechnology, 7,* 178–188.

Sato, R. (2014). *Bacillus thuringiensis* insecticidal toxins. In Insect Pathology (eds. Y. Kunimi and M. Kobayashi)., 35–49, Kodansya, Tokyo.

Schnepf, E., Crickmore, N., Van Rie, J., Lereclus, D., Baum, J., Feitelson, J., Zeigler, D. R. & Dean, D. H. (1998). *Bacillus thuringiensis* and its pesticidal crystal proteins. *Microbiology and Molecular Biology Reviews, 62,* 775–806.

Sher, D., Fishman, Y., Zhang, M., Lebendiker, M., Gaathon, A., Mancheño, J. M. & Zlotkin, E. (2005). Hydralysins, a new category of β-pore-forming toxins in Cnidaria. *Journal of Biological Chemistry, 280,* 22847–22855. doi: 10.1074/jbc.M503242200.

Shu, C., Su, H., Zhang, J., He, K., Huang, D. & Song, F. (2013). Characterization of cry9Da4, cry9Eb2, and cry9Ee1 genes from *Bacillus thuringiensis* strain T03B001. *Applied Microbiology and Biotechnology, 97,* 9705–9713. doi: 10.1007/s00253-013-4781-5.

Soberón, M., Lopez-Diaz, J. A. & Bravo, A. (2013). Cyt toxins produced by *Bacillus thuringiensis*: A protein fold conserved in several pathogenic microorganisms. *Peptides, 41*, 87–93. doi: 10.1016/ j.peptides. 2012.05.023.

Soberón, M., Pardo-López, L., López, I., Gómez, I., Tabashnik, B. E. & Bravo, A. (2007). Engineering modified Bt toxins to counter insect resistance. *Science, 318*, 1640–1642. doi: 10.1126/*science*.1146453.

Takemoto, Y., Mitsuhashi, W., Murakami, R., Konishi, H. & Miyamoto, K. (2008). The N-terminal region of an entomopoxvirus fusolin is essential for the enhancement of peroral infection, whereas the C-terminal region is eliminated in digestive juice. *Journal of Virology*, 82: 12406–12415. doi: 10.1128/JVI.01605-08.

Tabashnik, B. E., Finson, N., Groeters, F. R., Moar, W. J., Johnson, M. W., Luo, K. & Adang, M. J. (1994). Reversal of resistance to *Bacillus thuringiensis* in *Plutella xylostella* (Ietidal crysal protein/diamondback moth). *Proceedings of the National Academy of Sciences of the United States of America, 91*, 4120–4124. doi: 10.1073/pnas.91.10.4120.

Tabashnik, B. E., Fabrick, J. A., Unnithan, G. C., Yelich, A. J., Masson, L., Zhang, J. & Soberón, M. (2013). Efficacy of genetically modified Bt toxins alone and in combinations against pink bollworm resistant to Cry1Ac and Cry2Ab. *PloS One, 8*, 11, e80496. doi: 10.1371/journal.pone.0080496.

Takahashi, H., Nakaho, K., Ishihara, T., Ando, S., Wada, T., Kanayama, Y., Asano, S., Yoshida, S., Tsushima, S. & Hyakumachi, M. (2014). Transcriptional profile of tomato roots exhibiting *Bacillus thuringiensis*-induced resistance to *Ralstonia solanacearum. Plant Cell Reports, 33*, 99–110. doi: 10.1007/s00299-013-1515-1.

Torres-Quintero, M. C., Gómez, I., Pacheco, S., Sánchez, J., Flores, H., Osuna, J., Mendoza, G., Soberón, M. & Bravo, A. (2018). Engineering *Bacillus thuringiensis* Cyt1Aa toxin specificity from dipteran to lepidopteran toxicity. *Scientific Reports, 8*, 4989.125. doi: 10.1038/s41598-018-22740-9.

Walters, F. S., Stacy, C. M., Lee, M. K., Palekar, N. & Chen, J. S. (2008). An engineered chymotrypsin/cathepsin G site in Domain I renders

Bacillus thuringiensis Cry3A active against western corn rootworm larvae. *Applied and Environmental Microbiology, 74,* 367–374. doi: 10.1128/AEM.02165-07.

Wang, P. & Granados, R. R. (2001). Molecular structure of the peritrophic membrane (PM): Identification of potential PM target sites for insect control. *Archives of Insect Biochemistry and Physiology, 47,* 110–118. doi: 10.1002/arch.1041.

Wang, S., Kaina, W. & Wang, P. (2018). *Bacillus thuringiensis* Cry1A toxins exert toxicity by multiple pathways in insects. *Insect Biochemistry and Molecular Biology, 102,* 59–66. doi: 10.1016/j.ibmb.2018.09.013.

Wirth, M., Georghiou, G. P. & Federici, B. A. (1997). CytA enables CryIV endotoxins of *Bacillus thuringiensis* to overcome high levels of CryIV resistance in the mosquito *Culex quinquefasciatus. Proceedings of the National Academy of Sciences of the United States of America, 94,* 10536–10540. doi: 10.1073/pnas.94.20.10536.

Xiao, Y., Liu, K., Zhang, D., Gong, L., He, F., Soberón, M., Bravo, A., Tabashnik, B. E. & Wu, K. (2016). Resistance to *Bacillus thuringiensis* mediated by an ABC transporter mutation increases susceptibility to toxins from other bacteria in an invasive insect. *PLoS Pathogens, 12,* e1005450. doi: 10.1371/journal.ppat.1005450.

Zhang, S., Cheng, H., Gao, Y., Wang, G., Liang, G. & Wu, K. (2009). Mutation of an aminopeptidase N gene is associated with *Helicoverpa armigera* resistance to *Bacillus thuringiensis* Cry1Ac toxin. *Insect Biochemistry and Molecular Biology, 39,* 421–429. doi: 10.1016/j.ibmb.2009.04.003.

Zhang, X., Candas, M., Griko, N. B., Rose-Young, L. & Bulla, Jr. L. A. (2005). Cytotoxicity of *Bacillus thuringiensis* Cry1Ab toxin depends on specific binding of the toxin to the cadherin receptor BT-R1 expressed in insect cells. *Cell Death & Differentiation, 12,* 1407–1416. doi: 10.1038/sj.cdd.4401675.

Zhang, X., Candas, M., Griko, N. B., Taussig, R. & Bulla, Jr. L. A. (2006). A mechanism of cell death involving an adenylyl cyclase/PKA signaling pathway is induced by the Cry1AB toxin of *Bacillus thuringiensis.*

Proceedings of the National Academy of Sciences of the United States of America, 103, 9897–9902. doi: 10.1073/ pnas.0604017103.

Zhou, X., Ren, L., Meng, Q., Li, Y., Yu, Y. & Yu, J. (2010). The next-generation sequencing technology and application. *Protein & Cell, 1*, 520–536. doi: 10.1007/s13238-010-0065-3.

In: *Bacillus thuringiensis*
Editor: David P. Sanders

ISBN: 978-1-53619-570-5
© 2021 Nova Science Publishers, Inc.

Chapter 4

SOLUBILIZATION OF PHOSPHORUS BY *BACILLUS THURINGIENSIS*

Jorge Delfim[*]
Instituto de Investigação Agronômica, Chianga, Huambo, Angola
Department of Agronomy, State University of Londrina,
Londrina, Paraná, Brazil

ABSTRACT

Bacillus thuringiensis (*B. thuringiesis*) is widely known and used as a biopesticide in agriculture. It is used in the control of many crop pests. *B. thuringiensis* related studies are mostly focused on its insecticidal activity due to its entomopathogenic properties. Meanwhile, studies focusing on the biofertilizer or plant growth promoter features of *B. thuringiensis*, including its phosphorus (P) solubilizing capacity and interactions with plants, are limited. On the other hand, *B. thuringiensis* has the ability to solubilize organic and inorganic P forms in the soil and become available for plant growth. However, P is an essential nutrient for the plants and the second most important element after nitrogen. It is unavailable to plants because in the soil P is mostly present in the fixed form. *B. thruringiensis* having the phosphate solubilizing capacity, they convert the insoluble phosphate into soluble form through the changes the pH of medium,

---

[*] Corresponding Author's E-mail: jorgedelfim88@yahoo.com.

enzymes production, and organic acids and make it available for plant uptake and nutrition. It's a way and the opportunity to reduce/minimize the use of chemical fertilizers phosphate and to improve the availability of insoluble P forms in soil. In addition, the inoculation with beneficial microbes/bacteria such as *B. thruringiensis* is an ecological strategy and friendly to the environment that allows for involvement to the solubility of P in soil. Knowing the role of the *B. thuringiensis* for the crop production and environmental sustainability. The aims of this chapter are to revise, debate, and evaluate the effects of *B. thruringiensis* as phosphate solubilizing and P uptake by plant establishment.

Keywords: phosphorus solubilization, *Bacillus thuringiensis*, phosphorus in the soil and plant

INTRODUCTION

Phosphorus (P) available in several soils is very low, compared to the total soil P content. On the other hand, P is one of the principal macronutrients for plant growth and development. In this context, the scarce availability of P in soil solution has been causing the low yield of many crops. However, the problem of the low P solubility in the soil is principally by retention and fixation processes with the soil mineral compounds, P can be linked to aluminium (Al), iron (Fe), and calcium (Ca) according to the soil pH. Therefore, as a result of adsorption, precipitation and conversion to inorganic and organic forms in the soil. Also, its associated a largest P losses directly from farms through soil leaching and erosion (Elser and Bennett 2011), caused majority for the incorrect soil and crops management.

Agricultural soil P deficiency affect crop yield, notedly in tropical and subtropical regions of the world (Balemi and Negisho 2012). In actual agriculture, the application of mineral fertilizers is the main source of P used to crop production, principally in the developed countries, including in regions that remained in the mixed crop and livestock farming system (Le Noë et al. 2020).

Phosphate sources in the world are limited. Furthermore, the mineral P fertilizes applied every crop season in agriculture food production result in the accumulation of insoluble P in soils, so it can cause more losses of P in soil and water contamination (eutrophication). However, the use of phosphate solubilizing microorganisms such as *B. thuringiensis* can be an alternative to reduce the use, losses and P accumulation in soil and improve P acquisition for plant growth.

The objective of this chapter is to review, present, and discuss the use of some strains of the *B. thuringiensis* as phosphate solubilizing and can be used as a biofertilizer for crop production.

PHOSPHORUS IN THE SOIL

P is one of the principal nutrient for plants, animals, and microbes. In natural conditions, the availability of P in soil solution is low generally. However, an adequate supply of P is essential for all organisms due to its central role in biosynthesis and part of molecules (e.g., ATP and DNA), signaling (George et al. 2018). Phosphate rock is a principal source of P used in food production. P chemical fertilizers are produced generally using phosphate rock, and this is a non-renewable P source (Elser and Bennett 2011).

In soils, the P forms are inorganic (Pi) and organic (Po), and availability is low, principally due to the strong retention/fixation by the soil minerals such as clay, oxides, and hydroxides of Al and Fe. Po commonly present as soil microbial biomass and organic matter decomposition compounds (phosphomonoesters and diesters) (Nash et al. 2014). The form and compounds of the P which accumulate in soil depend on many factors such as soil type, pH and soil management. In acids soils, P availability is scarcer and heavily absorbed in the soil matrix. Po pools in the soil can represent a high quantity of the total soil P and its solubility is changeable and dependent on soil quality and quantity organic matter decomposition (Turner et al. 2015).

This has led to microorganisms such as *B. thuringiensis* developing a wide range of strategies to improve P availability in soil. Plants can only use Pi (i.e., HPO_4^{2-}, $H_2PO_4^-$), soil microorganisms can also potentially uptake Po and make it available in Pi form after the mineralization process (Nash et al. 2014).

In modern agriculture, P fertilizer applications greater than the quantity required by the plant are absorbed to minimize the strong bonding of the P to the soil minerals (MacDonald et al. 2011). This practice has greatly contributed to an increase in the P accumulation of cultivated soils, thus soils in several regions of the world have a high amount of insoluble P forms by crops. However, the accumulation of soil P also has negative consequences (financial loss to the farmer, environmental contaminations) (Elser and Bennett 2011).

Differents studies have been reported the accumulation of insoluble P forms in agricultural soils in several parts of the world (Haygarth et al. 2014; Le Noë et al. 2020; Pavinato et al. 2020), has become an important reservoir of P in much soils (Le Noë et al. 2020), the possible importance of the P accumulated in agricultural soils for supporting crop production and for a transition towards more sustainable agriculture needs to be further assessed.

In this case, the search for alternatives to minimize environmental problems, and increase the availability of P from the soil and its absorption by plants, the use of P-solubilizing microorganisms have shown potential for use in different cultures and, environments and regions as one friendly alternative to the environment.

The use of P-solubilizing microorganisms is a new strategy that can mobilize different P forms from the soil, increasing the P lability as well as decreased the no-labile P. Olso P-solubilizing microorganisms can prevent or reduce P accumulation and losses and help to have a substantial positive effect in soil and water conservation.

BACILLUS THURINGIENSIS

B. thuringiensis is a microorganism (bacteria) of the genus *Bacillus* includes strict aerobic or facultative anaerobic Gram-positive bacteria. *B. thuringiensis* form endospore, are resistant structures that can outlive adverse environmental conditions ((Delfim, Gerding, and Zagal 2020; Sauka et al. 2021).

B. *thuringiensis* is widely known for producing different substances that have beneficial effects and are used to control pests and diseases in crops. On the other hand, some strains of *B. thuringiensis* are also capable of solubilizing P, and enhancing plant growth and protection, the potential that these bacteria present in solubilizing large amounts of P has been studied in recent years (Wang, Liu, and Li 2014; Cherif-Silini et al. 2016; Xu et al. 2019; ALKahtani, Fouda, et al. 2020; Sauka et al. 2021) for the production of *B. thuringiensis*-based biofertilizers and other growth promotion products.

B. *thuringiensis* is largely distributed in soils throughout the world. It uses the soil as a natural habitat and *B. thuringiensis* spore cell, and the can survival in the soil under conditions of biotic or abiotic stress. Among the groups of microorganisms studied in relation to the production of bioproducts. In the group of bacteria, *B. thuringiensis* produce diverse microbial metabolites and some strains can solubilize phosphates (Cherif-Silini et al. 2016), (see, Table 1) and improve P acquisition for plants (see, Table 2).

MECHANISMS OF PHOSPHORUS SOLUBILIZATION BY *BACILLUS THURINGIENSIS*

P-solubilizing by *B. thuringiensis* involves different mechanisms (Table 1) such as activate indirect P solubilization by changing the pH of the medium, producing phosphatase enzymes, siderophores, excretion organic acids, production of phytases (Wang, Liu, and Li 2014; Cherif-Silini et al. 2016; Xu et al. 2019).

Change the pH

B. thuringiensis is considered rhizobacteria that acidified the pH of the medium (De Freitas, Banerjee, and Germida 1997), this strategy to change the pH in the rhizosphere can result in P solubilization under certain conditions.

The drop in medium pH is frequently reported along the P solubilization process the lowest pH values can result in the highest phosphate solubilization (Wang, Liu, and Li 2014; Xu et al. 2019; Sauka et al. 2021), and have been negative correlations between soluble phosphate concentration and pH in the mediums, this is attributed to the production of low molecular weight organic acids, which are the principal players to phosphate solubilization and drop of pH (Xu et al. 2019; Sauka et al. 2021).

Phosphatase and Phytase

The enzyme production is another way to P solubilization in several strains of *B. thuringiensis*. The Phosphatases classification depends based on the optimum pH required for their catalytic action, *B. thurigiensis* produce acidic, neutral, and alkaline organophosphorus phosphatases (Raddadi et al. 2008; Ambreen, Yasmin, and Aziz 2020), and production of phytases (Raddadi et al. 2008). However, phosphatase production is influenced by temperature and pH (Shahab and Ahmed, 2008), these enzymes dissolve phosphate according to the bacteria cell requirement for inorganic P using organophosphate and phosphonate sources. *B. thuringiensis* produce phosphatase for P solubilization (De Freitas, Banerjee, and Germida 1997; Delfim et al. 2018), acid and alkaline phosphatases catalyze the hydrolysis of organic P compounds when the level of inorganic P in the growth medium is limiting.

Table 1. *Bacillus thuringiensis* strains as phosphate solubilizing and other traits

B. thuringiensis (strains)	Isolated from	Substrats	Mode of action	Reference
MB497	-	Calcium phosphate	Phosphate solubilization, acid, neutral and alkaline phosphatase production.	(Ambreen, Yasmin, and Aziz 2020)
-	-	Rock phosphate ore	Phosphate solublization and indole acetic acid production	(Vassilev, Nikolaeva, and Vassileva 2007)
GL-1	-	Phosphate rock	Phosphate solubilization, gluconic, pyruvic, a-Ketoglutaric, citrate, and succinic acids production	(Xu et al. 2019)
CMG857	-	Zinc phosphate	Phosphate solubilization	(Shahab and Ahmed 2008)
GDB-1	Pinus sylvestris	Tricalcium phosphate	1-aminocyclopropane-1-carboxylic acid deaminase activity, indole acetic acid and siderophore production, and phosphate solubilization	(Babu, Kim, and Oh 2013)
SG1, SG6, SG11 and SG17	-	Tricalcium phosphate	Phosphate solubilization, catechol type siderophore production; hydroxamate type siderophore production, and indole acetic acid	(Lyngwi et al. 2016)
B1	Tea	Aluminum phosphate	Phosphate solubilization, and organic acid production	(Wang, Liu, and Li 2014)
B1, B4, B7, B8, B10, B18, B21, B23, BA8, BA11, BA14, BA15, D1, D4, D5, D12, and D14	Wheat	Tricalcium phosphate	Phosphate solubilization; siderophores, and indole acetic acid	(Cherif-Silini et al. 2016)
2P1M3	Pea	North Carolina rock phosphate	Phosphate solubilizetion, indole acetic acid, and acid and alkaline phosphatase activities	(De Freitas, Banerjee, and Germida 1997)
W29, and W30	Walnut	Tricalcium phosphate, aluminum phosphate and iron phosphate	Phosphate solubilization	(Yu et al. 2011)

                            Jorge Delfim

Table 1. (Continued)

B. thuringiensis (strains)	Isolated from	Substrats	Mode of action	Reference
SNKr10	Spinach	-	Phosphate solubilization, and 1-aminocyclopropane-1-carboxylic acid deaminase activity, ammonia production and biological nitrogen fixation	(Sharma and Saharan 2016)
C25	Lettuce	-	Phosphate solubilizetion, and indole acetic acid	(Gomes et al. 2003)
KVS25	Brassica juncea	Tricalcium phosphate	Phosphate solubilizetion, indole acetic acid, and hydrogen cyanide production	(Vishwakarma et al. 2018)
BNH2	Wheat	Tricalcium phosphate, zinc carbonate	Phosphate solubilization, zinc, and Siderophores	(Verma et al. 2016)
BtD5, BtI4 and BtI4	-	Tricalcium phosphate and rock phosphate	Phosphate solubilization, cellulase, chitinase, siderophore and indole acetic acid.	(Kassogué et al. 2016)
Fm.6	Fagonia mollis	Tricalcium phosphate	Phosphate solubilization, ammonia production, indole acetic acid production, amylase, pectinase, carboxymethyl cellulase, cellulase, xylanase and gelatinase	(ALKahtani, Fouda, et al. 2020)
X30	Amaranthus tricolor	-	Indole acetic acid and 1-aminocyclopropane-1-carboxylic acid deaminase activity	(Han, Wang, et al. 2018)
KKI-6 and JUKD-2	Soybean	Zinc oxide, zinc phosphate and zinc carbonate	Zinc solubilization and changes in pH	(Khande et al. 2017)
PG7, PG21, and PG31	-	-	Phosphate solubilization, ammonia production, indole acetic acid production, amylase, protease and cellulase activity.	(De Mandal, Singh, and Kumar 2018)

- not described or indicate

Organic Acids and Siderophores

B. thurigiensis produce organic acids, such as gluconic, 2-ketogluconic, citric, and oxalic acids (Xu et al. 2019), and the secretion of different organic acids by *B. thuringiensis* strains via the direct oxidation, which exhibited the ability to solubilize considerable amounts of tricalcium phosphate and others P sources (Sauka et al. 2021).

Wang, Liu, and Li (2014), observed, the correlation between pH value and the concentration of organic acid when *B thuringiensis B1* was inoculated, and this strain showed a P-solubilizing ability by producing a high concentration of organic acid. *B. thruringiensis* producre siderophores and it can improve plant growth and protect plants from several diseases due to competition effects for Fe (Raddadi et al. 2008; Cherif-Silini et al. 2016).

PHOSPHORUS SOLUBILIZATION BY *BACILLUS THURINGIENSIS* AND PLANT GROWTH

Solubilize phosphate is one of the positive benefices that some *B. thuringiensis* strains can provide for crops in several conditions (Table 2). Known the role of the P in plant growth and development, its a contribution of this bacterium in the food production process.

The inoculation with *B. thuringiensis* as a biofertilizer (i.e., phosphate solubilizing and plant growth) has been growing over time. In this context, De Freitas, Banerjee, and Germida (1997) inoculated *B. thuringiensis* in canola plants with North Carolina rock phosphate at a rate of 90.0 kg P ha^{-1} in a Dark Brown Chernozem soil and demonstrated that these bacteria have the potential to use as an inoculant for canola. On the other hand, Wang, Liu, and Li (2014) evaluated the P release efficiency of *B. thuringinsis* strain B1 inoculated in peanut growing in red clay soil (Ultisol), the Olsen-P in soil increased from 14.7 to 23.4 mg kg^{-1}, with solubilization of occluded P, and resulted in significantly increased plant P concentration and crude protein in grain content, they suggest the use of *B. thuringiensis* B1 as a biofertilizer.

 Jorge Delfim

Table 2. Attributes of *B. thuringiensis* as biofertilizer and plant growth promotion

B. thuringiensis strains	Route of promotion	Beneficial host	Growth conditions	Reference
Bt	Indole acetic acid, 1-aminocyclopropane-1-carboxylic acid deaminase production, phosphate solubilization, plant, and tolerate drought stress.	*Lavandula dentata*	Greenhouse	(Armada et al. 2016)
A5-BRSC	Plant growth promoting such as, seed germination, shoot height, root length, leaf diameter, vigor index, fruit weight, seed weight and total fresh weight, dry weight as well as nutritional content of the fruits.	*Abelmoschus esculentus*	Field	(Bandopadhyay 2020)
B1	Increase Olsen- P, plant total P concentration, dry weight in B1 treatment. plant height, number of branches and nodules per plant, plant heigh, plant hundred-seed weight and crude protein contente of seeds.	Peanut	-	(Wang, Liu, and Li 2014)
GL-1	Raise shoot biomass, photosynthetic rate, soil bioavailable P, and reduce the phytoavailability of Pb, Cd, and Zn in contaminated soils.	Lettuce	-	(Xu et al. 2019)
AG-82	Increase Olsen-P, acid phosphatase activity, soil microbial biomass, root biomass, P concentration in shoot and solubilizetion of P fraction extracted with conc. HCl-Po and HCl 1 mol in Andisol and Ultisol.	Wheat	Greenhouse	(Delfim et al. 2018; Delfim, Gerding, and Zagal 2020)
KVS25	Increase shhoot, length, root length, fresh weight organic carbon and, potassium in soil.	*Brassica juncea*	Growth chamber	(Vishwakarma et al. 2018)
-	Drought stress plants, increase N, C, P, K Mg, Ca, B, Fe, Zn and Cu acquisition.	Maize	Field	(Armada et al. 2015)
-	Improved Shoot and root weights, P, K, Ca, Mg, Zn and B plant content, decreased stomatal conductance and electrolyte leakage, high accumulation of proline in shoot and root, plants under drought stress conditions.	*Trifolium repens*	Field	(Ortiz et al. 2015)

B. thuringiensis strains	Route of promotion	Beneficial host	Growth conditions	Reference
-	Increase root colonization by G. mosseae and G. intraradice mycorrhizal, succinate dehydrogenase, alkaline phosphatase production, N and P content in plants under the lowest P conditions.	Lettuce	Growth chamber	(Vivas et al. 2003)
BtD5, BtI4 and BtI4	Increased % germination, root length, shoot length, seedling length and vigor index.	Maize	-	(Kassogué et al. 2016)
Fm.6	Increased Dry Weight of shoot Nutrients Content in shoot (P, N, and K)	Maize	Greenhouse	(ALKahtani, Fouda, et al. 2020)
MH161336	Regulation of proline accumulation, increased chlorophyll fluorescence, decreased lipid peroxidation, superoxide, hydrogen peroxide, catalase, peroxidase ctivity, superoxide dismutase and glutathione reductase enzymes.	Sweet pepper	-	(ALKahtani, Attia, et al. 2020)
	increased, number of fruits, fruit fresh weight and total fruit yield under salinity stress conditions.			
X30	Increase polyamine production and pH in solution. immobilize Cd, reduce the available Cd and its effect in plant growth, produce arginine decarboxylase, and increase root and seed.	*Brassica napus*	Greenhouse	(Han, Wang, et al. 2018)
X30	Increase pH in solution, immobilize Cd and Pb, increase plant growth, reduce Cd and Pb uptake, and alleviate Cd and Pb toxicity/stress to the plant	*Raphanus raphanistrum*	Field	(Han, Sheng, et al. 2018)
KKI-6 and JUKD-2	Increase zinc availability in soil, and zinc content in seed number and dry weight of nudule and root, shoot, straw and seed weight in soybean, and number of tillers, and seeds and straw weight of wheat	Soybean and wheat	Microcosm	(Khande et al. 2017)

- not described or indicate

Armada et al. (2015), reported a significant increase in P acquisition by maiz plants under drought condition inoculated with *B. thuringiensis* alone and co-inoculation of *B. thuringiensis* + *Arbuscular mycorrhizal*. The results showed that microbial treatments were the principal responsible for P nutrition and plant growth under stress conditions. While, Mishra et al.

(2009), evaluated the effects of co-inoculation of *B. thuringiensis*-KR1 and *Bradyrhizobium japonicum*-SB1 on the growth of soybean, the co-inoculation resulted in an increase in nodule number, root biomass, root length, and shoot biomass in soybean plants, and confirming that *B. thuringiensis*-KR1 did not produce any antibacterial compound against *Bradyrhizobium japonicum* SB1. Same *B. thuringiensis* strains produce diverse organic acid, other compounds, and P solubilization (Table 1) that can help the plant growth and development. Besides, *B. thuringiensis*-KR1 was positively affected by pea and lentil increased nodule number and root biomass (Mishra, Mishra, Selvakumar, Bisht, et al. 2009)

Bandopadhyay (2020), evaluated *B. thuringiensis* strain A5-BRSC, applied as charcoalbased biofertilizer on *Abelmoschus esculentus* plants, the inoculation increased 75% phosphorus content in pods and improved the yield components of plants. Also, the strain A5-BRSC improved P and leaf protein content in *Amaranthus viridis, Capsicum annuum*, and *Ocimum tenuiflorum* crops (Bandopadhyay and Gangopadhyay. 2015).

Delfim et al. (2018) report that *B trhuringiensis* inoculation increased P-Olsen in the rhizosphere of wheat plants growing in an Ultisol and enhanced P concentration in plant tissues with the inoculation. The important role that *B. thuringiensis* play in plant nutrition by increasing available P in the soil can be attributed to the capacity that *B. thuringiensis* have to solubilize insoluble P forms in soil and different P fractions extracted in conc. HCl - Po and HCl 1 mol Pi (Delfim, Gerding, and Zagal 2020). Furthermore, these microbes solubilize P bonding in soil (Delfim, Gerding, and Zagal 2020) and recycling is important to the availability of soil, plant nutrition, and reduce P losses, water contamination, and P legacy in diverse agriculture soils, principally countries that have intense agriculture exploration and apply the high quantity of P fertilizers in the soil.

The co-inoculation of the *B. thuringiensis* with other microorganisms is one characteristic that can be used to produce biofertilizer or biostimulate for plants. Therefore, the co-inoculation, such as *B. thuringiensis* with *mycorrhizal fungi* have been reported that enhance plant growth promotion and P nutrition in different crops, principally under stress or limited conditions the microbial interactions, like the one's *B. thuringiensis* strains, play an important role in mycorrhizal vitality and thus, on plant survival ((Vivas et al. 2003; Armada et al. 2015; Armada et al. 2016).

Also, the inoculation of *B. thuringiensis* alone or with other products, such as chitosan application can improve plant growth under salinity stress condition due to the formation of growth regulators, solubilization and fixation of nutrients, and other physiological and biochemical strategies that can to cope with the deleterious effects of salinity on crops (ALKahtani, Attia, et al. 2020).

CONCLUSION

B thuringiensis is widely known as a biocontrol capacity for pests and diseases. however, some strains of *B. thuringiensis* can solubilize differents P forms in soil and make it available by crops. *B. thuringiensis* transform insoluble P into soluble forms using different mechanisms such as drop the pH of the medium, organic acid production, excreting phosphatases, and other compounds.

The success of the P solubilization process depends, the *B. thuringiensis* strains, soil type, management, environmental conditions, indigenous microbes in the soil, and the plant species. On the other hand, the quantity of P solubilization, mobilization, or mineralization by *B. thuringiensis* can be influenced principally by the P sources and structure of chemical forms in the soil.

The presence of large quantities of clay minerals associated with Fe and Al on soil surfaces can be resulting in very high P fixation capacity with very stable residual P pools that can difficult the microbial attack and consequently the P solubilizing process by *.B thuringiensis* and less P availability in the soil and acquisition for the plants.

B. thuringiensis solubilize insoluble phosphate compounds in soils and present potential to be used as a biofertilizer and produce diverse beneficial compounds helpful for plant growth. Therefore, this other characteristic of *B. thuringiensis* increases the are role in sustainable crop production and for global food security. Furthermore, some strains of *B. thuringiensis*, largely identified can be used as biocontrol, biofertilizer, and plant growth promoters. Also, the co-inoculation (mix) with *B. thuringiensis* and other microbes such as fungal and bacteria have been tested in several crops and environmental conditions are other approaches that be helpful and to improve the benefits for plants and *B. thuringiensis* can result in a potential future biofertilizer or stimulator product for sustainable agriculture to use on crop production.

REFERENCES

ALKahtani, Muneera D. F., Kotb A. Attia, Yaser M. Hafez, Naeem Khan, Ahmed M. Eid, Mohamed A. M. Ali. & Khaled A. A. Abdelaal. (2020). "Chlorophyll Fluorescence Parameters and Antioxidant Defense System Can Display Salt Tolerance of Salt Acclimated Sweet Pepper Plants Treated with Chitosan and Plant Growth Promoting Rhizobacteria." *Agronomy*, *10* (8), 1180. https://doi.org/10.3390/ agronomy10081180.

ALKahtani, Muneera D. F., Amr Fouda, Kotb A. Attia, Fahad Al-Otaibi, Ahmed M. Eid, Emad El-Din Ewais, Mohamed Hijri., et al. (2020). "Isolation and Characterization of Plant Growth Promoting Endophytic Bacteria from Desert Plants and Their Application as Bioinoculants for Sustainable Agriculture." *Agronomy*, *10* (9), 1–18. https://doi.org/ 10.3390/agronomy10091325.

Ambreen, Samina, Azra Yasmin. & Satara Aziz. (2020). "Isolation and Characterization of Organophosphorus Phosphatases from Bacillus Thuringiensis MB497 Capable of Degrading Chlorpyrifos, Triazophos and Dimethoate." *Heliyon*, *6* (7), e04221. https://doi.org/10.1016/ j.heliyon.2020.e04221.

Armada, E., Probanza, A., Roldán, A. & Azcón, R. (2016). "Native Plant Growth Promoting Bacteria Bacillus Thuringiensis and Mixed or Individual Mycorrhizal Species Improved Drought Tolerance and Oxidative Metabolism in Lavandula Dentata Plants." *Journal of Plant Physiology*, *192*, 1–12. https://doi.org/10.1016/j.jplph.2015.11.007.

Armada, Elisabeth, Rosario Azcón, Olga M. López-Castillo, Mónica Calvo-Polanco. & Juan Manuel Ruiz-Lozano. (2015). "Autochthonous Arbuscular Mycorrhizal Fungi and Bacillus Thuringiensis from a Degraded Mediterranean Area Can Be Used to Improve Physiological Traits and Performance of a Plant of Agronomic Interest under Drought Conditions." *Plant Physiology and Biochemistry*, *90*, 64–74. https://doi.org/10.1016/j.plaphy.2015.03.004.

Babu, A. Giridhar, Jong Dae Kim. & Byung Taek Oh. (2013). "Enhancement of Heavy Metal Phytoremediation by Alnus Firma with Endophytic Bacillus Thuringiensis GDB-1." *Journal of Hazardous Materials*, *250–251*, 477–83. https://doi.org/10.1016/ j.jhazmat. 2013.02.014.

Balemi, T. & Negisho, K. (2012). "Management of Soil Phosphorus and Plant Adaptation Mechanisms to Phosphorus Stress for Sustainable Crop Production: A Review." *Journal of Soil Science and Plant Nutrition*, *12* (3), 547–61. https://doi.org/10.4067/s0718-9516201200 5000015.

Bandopadhyay, Sandip. & Swati R. Gangopadhyay. (2015). "Studies on the effect of biofertilizer comprising of the plant growth promoting rhizobacteria Bacillus thuringiensis on different types of plants." *Prajnan O Sadhona – A Science Annual*, (2), 72-86.

Bandopadhyay, Sandip. (2020). "Application of Plant Growth Promoting Bacillus Thuringiensis as Biofertilizer on Abelmoschus Esculentus Plants under Field Condition." *Journal of Pure and Applied Microbiology*, *14* (2), 1287–94. https://doi.org/10.22207/ JPAM.14.2.24.

Cherif-Silini, Hafsa, Allaoua Silini, Bilal Yahiaoui, Imen Ouzari. & Abdellatif Boudabous. (2016). "Phylogenetic and Plant-Growth-Promoting Characteristics of Bacillus Isolated from the Wheat Rhizosphere." *Annals of Microbiology*, *66* (3), 1087–97. https://doi.org/10.1007/s13213-016-1194-6.

Delfim, Jorge, Macarena Gerding. & Erick Zagal. (2020). "Phosphorus Fractions in Andisol and Ultisol Inoculated with Bacillus Thuringiensis and Phosphorus Uptake by Wheat." *Journal of Plant Nutrition, 43* (18), 1–12. https://doi.org/10.1080/ 01904167.2020.1793176.

Delfim, Jorge, Mauricio Schoebitz, Leandro Paulino, Juan Hirzel. & Erick Zagal. (2018). "Phosphorus Availability in Wheat, in Volcanic Soils Inoculated with Phosphate-Solubilizing Bacillus Thuringiensis." *Sustainability (Switzerland), 10* (1). https://doi.org/10.3390/su100 10144.

Elser, James. & Elena Bennett. (2011). "Phosphorus Cycle: A Broken Biogeochemical Cycle." *Nature.* Nature Publishing Group. https://doi.org/10.1038/478029a.

Freitas, J. R. De, Banerjee, M. R. & Germida, J. J. (1997). "Phosphate-Solubilizing Rhizobacteria Enhance the Growth and Yield but Not Phosphorus Uptake of Canola (Brassica Napus L.)." *Biology and Fertility of Soils, 24* (4), 358–64. https://doi.org/10.1007/ s003740050258.

George, T. S., Giles, C. D., Menezes-Blackburn, D., Condron, L. M., Gama-Rodrigues, A. C., Jaisi, D., Lang, F., et al. (2018). "Correction to: Organic Phosphorus in the Terrestrial Environment: A Perspective on the State of the Art and Future Priorities (Plant and Soil, (2018), 427, 1-2, (191-208), 10.1007/S11104-017-3391-X)." *Plant and Soil, 427* (1–2), 209–11. https://doi.org/10.1007/s11104-017-3488-2.

Gomes, Andréa M. A., Rosa L. R. Mariano, Elineide B. Silveira. & Júlio C. P. Mesquita. (2003). "Isolamento, Seleção de Bactérias e Efeito de Bacillus Spp. Na Produção de Mudas Orgânicas de Alface." *Horticultura Brasileira, 21* (4), 699–703. https://doi.org/10.1590/ s0102-05362003000400026. ["Isolation, Selection of Bacteria and Effect of Bacillus Spp. In the Production of Organic Lettuce Seedlings." *Brazilian Horticulture*]

Han, Hui, Xiafang Sheng, Jingwen Hu, Linyan He. & Qi Wang. (2018). "Metal-Immobilizing Serratia Liquefaciens CL-1 and Bacillus Thuringiensis X30 Increase Biomass and Reduce Heavy Metal Accumulation of Radish under Field Conditions." *Ecotoxicology and Environmental Safety, 161* (October), 526–33. https://doi.org/10.1016/ j.ecoenv.2018.06.033.

Han, Hui, Qi Wang, Lin yan He. & Xia fang Sheng. (2018). "Increased Biomass and Reduced Rapeseed Cd Accumulation of Oilseed Rape in the Presence of Cd-Immobilizing and Polyamine-Producing Bacteria." *Journal of Hazardous Materials, 353* (July), 280–89. https://doi.org/ 10.1016/j.jhazmat.2018.04.024.

Haygarth, Philip M., Helen P. Jarvie, Steve M. Powers, Andrew N. Sharpley, James J. Elser, Jianbo Shen, Heidi M. Peterson., et al. (2014). "Sustainable Phosphorus Management and the Need for a Long-Term Perspective: The Legacy Hypothesis." *Environmental Science and Technology.* American Chemical Society. https://doi.org/10.1021/ es502852s.

Kassogué, Adounigna, Amadou Dicko, Diakaridia Traoré, Rokiatou Fané, Fernando Valicente. & Amadou Babana. (2016). "Bacillus Thuringiensis Strains Isolated from Agricultural Soils in Mali Tested for Their Potentiality on Plant Growth Promoting Traits." *British Microbiology Research Journal, 14* (3), 1–7. https://doi.org/ 10.9734/bmrj/2016/24785.

Khande, Rajesh, Sushil K. Sharma, Aketi Ramesh. & Mahaveer P. Sharma. (2017). "Zinc Solubilizing Bacillus Strains That Modulate Growth, Yield and Zinc Biofortification of Soybean and Wheat." *Rhizosphere, 4* (December), 126–38. https://doi.org/10.1016/ j.rhisph.2017.09.002.

Lyngwi, Nathaniel A., Macmillan Nongkhlaw, Debajit Kalita. & Santa Ram Joshi. (2016). "Bioprospecting of Plant Growth Promoting Bacilli and Related Genera Prevalent in Soils of Pristine Sacred Groves: Biochemical and Molecular Approach." *PLoS ONE, 11* (4), 1–13. https://doi.org/10.1371/journal.pone.0152951.

MacDonald, Graham K., Elena M. Bennett, Philip A. Potter. & Navin Ramankutty. (2011). "Agronomic Phosphorus Imbalances across the World's Croplands." *Proceedings of the National Academy of Sciences of the United States of America, 108* (7), 3086–91. https://doi.org/ 10.1073/pnas.1010808108.

Mandal, Surajit De, Sambanduram Samarjit Singh. & Nachimuthu Senthil Kumar. (2018). "Analyzing Plant Growth Promoting Bacillus Sp. and Related Genera in Mizoram, Indo-Burma Biodiversity Hotspot."

Biocatalysis and Agricultural Biotechnology, 15 (July), 370–76. https://doi.org/10.1016/j.bcab.2018.07.026.

Mishra, Pankaj K., Smita Mishra, Selvakumar, G., Bisht, J. K., Samresh Kundu. & Hari Shankar Gupta. (2009). "Coinoculation of Bacillus Thuringeinsis-KR1 with Rhizobium Leguminosarum Enhances Plant Growth and Nodulation of Pea (Pisum Sativum L.) and Lentil (Lens Culinaris L.)." *World Journal of Microbiology and Biotechnology* 25 (5): 753–61. https://doi.org/10.1007/s11274-009-9963-z.

Mishra, Pankaj K., Smita Mishra, G. Selvakumar, Samresh Kundu. & Hari Shankar Gupta. (2009). "Enhanced Soybean (Glycine Max L.) Plant Growth and Nodulation by Bradyrhizobium Japonicum-SB1 in Presence of Bacillus Thuringiensis-KR1." *Acta Agriculturae Scandinavica Section B: Soil and Plant Science*, 59 (2), 189–96. https://doi.org/10.1080/09064710802040558.

Nash, David M., Philip M. Haygarth, Benjamin L. Turner, Leo M. Condron, Richard W. McDowell, Alan E. Richardson, Mark Watkins. & Michael W. Heaven. (2014). "Using Organic Phosphorus to Sustain Pasture Productivity: A Perspective." *Geoderma*, 221–222, 11–19. https://doi.org/10.1016/j.geoderma.2013.12.004.

Noë, J. Le., Roux, N., Billen, G., Gingrich, S., Erb, K. H., Krausmann, F., Thieu, V., Silvestre, M. & Garnier, J. (2020). "The Phosphorus Legacy Offers Opportunities for Agro-Ecological Transition (France 1850-2075)." *Environmental Research Letters*, 15 (6). https://doi.org/10.1088/1748-9326/ab82cc.

Ortiz, N., Armada, E., Duque, E., Roldán, A. & Azcón, R. (2015). "Contribution of Arbuscular Mycorrhizal Fungi and/or Bacteria to Enhancing Plant Drought Tolerance under Natural Soil Conditions: Effectiveness of Autochthonous or Allochthonous Strains." *Journal of Plant Physiology*, 174, 87–96. https://doi.org/10.1016/ j.jplph. 2014.08.019.

Pavinato, Paulo S., Maurício R. Cherubin, Amin Soltangheisi, Gustavo C. Rocha, Dave R. Chadwick. & Davey L. Jones. (2020). "Revealing Soil Legacy Phosphorus to Promote Sustainable Agriculture in Brazil."

Scientific Reports, 10 (1), 1–11. https://doi.org/10.1038/s41598-020-72302-1.

Raddadi, Noura, Ameur Cherif, Abdellatif Boudabous. & Daniele Daffonchio. (2008). "Screening of Plant Growth Promoting Traits of Bacillus Thuringiensis." *Annals of Microbiology, 58* (1), 47–52. https://doi.org/10.1007/BF03179444.

Sauka, Diego Herman, Carlos Fabián Piccinetti, Daniela Adriana Vallejo, María Inés Onco, Melisa Paula Pérez. & Graciela Beatriz Benintende. (2021). "New Entomopathogenic Strain of Bacillus Thuringiensis Is Able to Solubilize Different Sources of Inorganic Phosphates." *Applied Soil Ecology, 160,* (November 2020). https://doi.org/10.1016/j.apsoil.2020.103839.

Shahab, Sadaf. & Nuzhat Ahmed. (2008). "Effect of Various Parameters on the Efficiency of Zinc Phosphate Solubilization by Indigenous Bacterial Isolates." *African Journal of Biotechnology, 7* (10), 1543–49. https://doi.org/10.5897/AJB07.145.

Sharma, Nisha. & Baljeet Saharan. (2016). "Bacterization Effect of Culture Containing 1-Aminocyclopropane-1-Carboxylic Acid Deaminase Activity Implicated for Plant Development." *British Microbiology Research Journal, 16* (1), 1–10. https://doi.org/10.9734/bmrj/2016/27135.

Turner, Benjamin L., Alexander W. Cheesman, Leo M. Condron, Kasper Reitzel. & Alan E. Richardson. (2015). "Introduction to the Special Issue: Developments in Soil Organic Phosphorus Cycling in Natural and Agricultural Ecosystems." *Geoderma,* 257–258, 1–3. https://doi.org/10.1016/j.geoderma.2015.06.008.

Vassilev, Nikolay, Iana Nikolaeva. & Maria Vassileva. (2007). "Indole-3-Acetic Acid Production by Gel-Entrapped Bacillus Thuringiensis in the Presence of Rock Phosphate Ore." *Chemical Engineering Communications, 194* (4), 441–45. https://doi.org/10.1080/00986440600983486.

Verma, Priyanka, Ajar Nath Yadav, Kazy Sufia Khannam, Sanjay Kumar, Anil Kumar Saxena. & Archna Suman. (2016). "Molecular Diversity and Multifarious Plant Growth Promoting Attributes of Bacilli

Associated with Wheat (*Triticum Aestivum* L.) Rhizosphere from Six Diverse Agro-Ecological Zones of India." *Journal of Basic Microbiology*, *56* (1), 44–58. https://doi.org/10.1002/jobm.201500459.

Vishwakarma, Kanchan, Vivek Kumar, Durgesh K. Tripathi. & Shivesh Sharma. (2018). "Characterization of Rhizobacterial Isolates from Brassica Juncea for Multitrait Plant Growth Promotion and Their Viability Studies on Carriers." *Environmental Sustainability*, *1* (3), 253–65. https://doi.org/10.1007/s42398-018-0026-y.

Vivas, A., Marulanda, A., Gómez, M., Barea, J. M. & Azcón, R. (2003). "Physiological Characteristics (SDH and ALP Activities) of Arbuscular Mycorrhizal Colonization as Affected by *Bacillus Thuringiensis* Inoculation under Two Phosphorus Levels." *Soil Biology and Biochemistry*, *35* (7), 987–96. https://doi.org/10.1016/S0038-0717(03)00161-5.

Wang, Tong, Man Qiang Liu. & Hui Xin Li. (2014). "Inoculation of Phosphate-Solubilizing Bacteria *Bacillus Thuringiensis* B1 Increases Available Phosphorus and Growth of Peanut in Acidic Soil." *Acta Agriculturae Scandinavica Section B: Soil and Plant Science*, *64* (3), 252–59. https://doi.org/10.1080/09064710.2014.905624.

Xu, Jia cheng, Li min Huang, Chengyu Chen, Jing Wang. & Xin xian Long. (2019). "Effective Lead Immobilization by Phosphate Rock Solubilization Mediated by Phosphate Rock Amendment and Phosphate Solubilizing Bacteria." *Chemosphere*, *237*, 124540. https://doi.org/10.1016/j.chemosphere.2019.124540.

Yu, Xuan, Xu Liu, Tian Hui Zhu, Guang Hai Liu. & Cui Mao. (2011). "Isolation and Characterization of Phosphate-Solubilizing Bacteria from Walnut and Their Effect on Growth and Phosphorus Mobilization." *Biology and Fertility of Soils*, *47* (4), 437–46. https://doi.org/10.1007/s00374-011-0548-2.

INDEX